U0875303

普通高等教育“十一五”国家级规划教材
全国高等医药院校药学类实验双语教材

微生物学实验与指导

（第二版）

Experiment and Guide for Microbiology

（Second Edition）

主　编　周长林
副主编　陈向东　徐旭东　苏　昕
编　委　（以姓氏笔画为序）
马晓楠（沈阳药科大学）
张文军（广东药学院）
苏　昕（沈阳药科大学）
周长林（中国药科大学）
陈向东（中国药科大学）
徐旭东（东南大学）
窦　洁（中国药科大学）

中国医药科技出版社

内容提要

本书是全国高等医药院校药学类实验双语教材之一。全书共分八个部分，内容主要包括微生物学实验基本原理、基本方法，细菌、放线菌、真菌、病毒等的形态结构和培养特征，微生物遗传学、免疫学和药学微生物学实验技术。本书可供高等医药院校药学类专业微生物学实验教学使用，同时可供药品质量检验人员参考。

图书在版编目（CIP）数据

微生物学实验与指导/周长林主编．—2版．—北京：中国医药科技出版社，2010.1

全国高等医药院校药学类实验双语教材．普通高等教育"十一五"国家级规划教材

ISBN 978-7-5067-4331-0

Ⅰ.微…　Ⅱ.周…　Ⅲ.微生物学-实验-双语教学-医学院校-教材　Ⅳ.Q93-33

中国版本图书馆CIP数据核字（2009）第142506号

美术编辑　陈君杞
版式设计　郭小平

出版　中国医药科技出版社
地址　北京市海淀区文慧园北路甲22号
邮编　100082
电话　发行：010-62227427　邮购：010-62236938
网址　www：cmstp.com
规格　787×1092mm 1/16
印张　22 3/4
字数　493千字
初版　2004年1月第1版
版次　2010年1月第2版
印次　2013年1月第2版第6次印刷
印刷　北京地泰德印刷有限责任公司
经销　全国各地新华书店
书号　ISBN 978-7-5067-4331-0
定价　42.00元

全国高等医药院校药学类规划教材常务编委会

出版说明

全国高等医药院校药学类规划教材是目前国内体系最完整、专业覆盖最全面、作者队伍最权威的药学类教材。随着我国药学教育事业的快速发展，药学及相关专业办学规模和水平的不断扩大和提高，课程设置的不断更新，对药学类教材的质量提出了更高的要求。

全国高等医药院校药学类规划教材编写委员会在调查和总结上轮药学类规划教材质量和使用情况的基础上，经过审议和规划，组织中国药科大学、沈阳药科大学、广东药学院、北京大学药学院、复旦大学药学院、四川大学华西药学院、北京中医药大学、西安交通大学医学院、华中科技大学同济药学院、山东大学药学院、山西医科大学药学院、第二军医大学药学院、山东中医药大学、上海中医药大学和江西中医学院等数十所院校的教师共同进行药学类第三轮规划教材的编写修订工作。

药学类第三轮规划教材的编写修订，坚持紧扣药学类专业本科教育培养目标，参考执业药师资格准入标准，强调药学特色鲜明，体现现代医药科技水平，进一步提高教材水平和质量。同时，针对学生自学、复习、考试等需要，紧扣主干教材内容，新编了相应的学习指导与习题集等配套教材。

本套教材由中国医药科技出版社出版，供全国高等医药院校药学类及相关专业使用。其中包括理论课教材82种，实验课教材38种，配套教材10种，其中有45种入选普通高等教育“十一五”国家级规划教材。

全国高等医药院校药学类规划教材

编写委员会

2009年8月1日

序

实验教学是高等药学院校最基本的教学形式之一，对培养学生科学的思维与方法、创新意识与能力，全面推进素质教育有着重要的作用。飞速发展的科学技术，已成为主导社会进步的重要因素。高等药学院校必须不断更新教学内容，以学科发展的前沿知识充实实验课程内容。

近年来，中国药科大学坚持以研究促教改，通过承担教育部“世行贷款——21世纪初高等教育教学改革项目”及立项校内教改课题等多种方式，调动了广大教师投身教学改革的积极性，将转变教师的教育思想观念与教学内容、教学方法的改革紧密结合起来，取得了实效。此次推出的国家“十一五”规划教材——药学专业双语实验教学系列，是广大教师长期钻研实验课程教学体系，改革教学内容，实现教育创新的重要成果。他们站在21世纪教育、科技和社会发展趋势的高度，对药学专业实验课程的教学内容进行了“精选”、“整合”和“创新”，强调对学生的动手能力、创新思维、科学素养等综合素质的全面培养。这套教材具有以下的特点：

1．教材将各学科的实验内容进行了广泛的“精选”，既体现了高等药学教育“面向世界、面向未来、面向现代化”，也考虑到我国药学教育的现状与实际；既体现了各门实验课程自身的独立性、系统性和科学性，又充分考虑到各门实验课程之间的联系与衔接，有助于学生在教学大纲规定的实验教学学时内掌握基本操作技术，提高动手能力，养成严谨、求实、创新的科学态度。

2．教材中新增的综合性、设计性实验有利于学生全面了解和综合掌握本门实验课程的教学内容。这一举措既满足了学生个性发展的需要，更注重培养学生分析问题、解决问题的能力和创新意识。

3．教材中适当安排一些反映药学学科发展前沿的实验，有利于学生在掌握实验基本技术的同时，对药学学科的新进展、新技术有所了解，激发他们学习药学知识与相关学科的兴趣。

4．教材以实践教学为突破口，采用双语体系编写，为实验课程改革构建数字化、信息化和外语教学的平台，有利于提高学生的科技英语水平。通过我校多年的药学系列实验课程双语教学实践，证明学生完全能够接受此套教材的教学。

国家十一五规划教材——药学专业双语实验教学系列教材的陆续出版，必将对推动我国高等药学教育的健康发展，产生积极而深远的影响。由于采用双语体系编写药学教学实验丛书尚属首次，缺乏经验，在内容选择及编写方法上的不妥之处，在所难免。欢迎从事药学教育的同行们批评赐教。

（中国药科大学校长、博士、教授、博士生导师）

Preface

Experimental teaching is one of the most fundamental teaching means in pharmaceutical colleges, playing an important role in training scientific thoughts and methods, creative consciousness and ability of the students as well as in promoting quality-oriented education in all-round way. Fast-advancing science and technology has come to be an important factor in dominating social progress. Teaching materials must be updated continually in pharmaceutical colleges, especially enriching the materials of experimental courses with the most advanced knowledge in the subject.

In recent years, China Pharmaceutical University have been stressing the promotion of teaching reform on the basis of research, succeeding in stimulating teachers' enthusiasm for teaching reform by various means such as undertaking the project of teaching reform in higher education at the beginning of 21st century sponsored financially by World Bank and entrusted by the Ministry of Education as well as approving and ratifying internal programs on teaching reform. Meanwhile, it yields fruits to integrate the transforming of teachers' educational ideology into the reform of teaching materials and methods. This series of textbook of national "the Eleventh Five-Year Plan" bilingual pharmaceutical experimental teaching series, is an important achievement made through studying ueaching system of experimental courses for long, reforming teaching materials and carrying out educational innovation of all the teachers concerned.

Meeting the new demands for education, science and technology and social growth, they select, integrate and innovate the teaching materials of pharmaceutical experimental courses, stressing the overall cultivation of comprehensive qualities, including experimental ability, creative thought and scientific attainments. This set of textbook possesses the following features:

1. These textbooks make an extensive "selection" of the experimental materials of each subject, reflecting the goal of facing the world, facing the future and facing the modernization in higher pharmaceutical education, and taking into account the status quota and reality of our pharmaceutical education; meanwhile embodying the individuality, systematicness and scientificalness of each experimental courses, which helps the students to grasp basic techniques of operation within the class hours of experimental teaching pre-

scribed by teaching syllabus and to improve their experimental ability and finally to cultivate a scientific approach of precision, practicality and creation.

2. The comprehensive designing experiments newly supplemented in the textbooks help the students to learn totally and grasp comprehensively the teaching materials of the experimental courses, which not only meets the students' needs for individual development but also trains their ability to analyze and solve problems and cultivates their creative consciousness.

3. Some experiments representing the latest development in pharmacy are properly included in the textbooks, which helps the students to learn about new advance and technology in pharmacy and to further arouse their interests in studying pharmacy and relevant subjects while grasping some basic techniques of experiment.

4. The textbooks take experimental teaching as starting point and are compiled in a system of bilingualism and aim to set up a platform of digitalization, information and foreign language teaching for the purpose of reforming experimental courses, which serves to enhance the students' level of technological English. It has been proved that the students have no difficulty being adapted to the teaching of this set of textbook through many years of bilingual teaching practice carried out in a series of pharmaceutical experimental courses of our university.

The successive publishing of the series of textbooks used for bilingual pharmaceutical experimental teaching-the national "the Eleventh Five-Year Plan" textbooks, will surely produce good and far-reaching influence in promoting the sound development of higher pharmaceutical education of our country. Since it is the first time that we have compiled this series of textbook of pharmaceutical teaching experiment in a bilingual system, we lack experience and thus some defects in choice of materials and way of compilation are inevitable. Experts engaged in pharmaceutical education are welcome to give any criticisms and advice.

Wu Xiaoming

Ph. D, prof., and supervisor of doctoral candidates

President of China Pharmaceutical University

第二版前言

本书是全国高等医药院校药学类实验双语教材之一，供药学专业微生物学实验教学使用。全书共分八个部分，内容主要包括微生物学实验基本原理、基本方法，细菌、放线菌、真菌、病毒等的形态结构和培养特征，微生物遗传学、免疫学和药学微生物学实验技术。其中第一部分由马晓楠老师编写，第二部分由周长林老师编写，第三部分由马晓楠和苏昕老师编写，第四部分由徐旭东老师编写，第五部分由张文军老师编写，第六部分由陈向东老师编写，第七部分由窦洁老师编写，第八部分由苏昕和窦洁老师编写。

实验编排一般包括实验目的、实验原理、实验材料、实验方法、结果与讨论五部分，力求使学生在实验中有明确的实验思路。结果和讨论部分旨在培养学生分析问题和全面总结实验结果的能力，培养独立思考的习惯，以期进一步消化和巩固所学的理论知识。实验所需的染料、培养基配方、试剂及缓冲液配制和微生物学实验的英文词汇列于附录中。

本书实验内容可根据教学需要适当选做，同时可供药物质检人员参考。

本书是编写教师多年教学实践的结晶，具有药学专业微生物学实验教学的显著特色，期望在药学生的素质教育方面能尽微薄之力。由于编者水平有限，在实验内容编排和中英文表述方面定有欠妥之处，恳请指正。

编　者

2009 年 8 月

Preface to the Second Edition

This text is a part of the series of pharmaceutical textbooks and designed to be an effective teaching tool for pharmaceutical university students. It is divided into eight parts which are organized by the principles of microbiological experiment and primary methods, the morphology and culture characteristics of bacteria, actinomycetes, fungi and virus, as well as the experimental technology of immunology and pharmaceutical microbiology. In the text, part one was written by Xiaonan Ma, part two was written by Changlin Zhou, part three was written by Xiaonan Ma and Xin Su, part four was written by Xudong Xu, part five was written by Wenjun Zhang, part six was written by Xiangdong Chen, part seven was written by Jie Dou, and part eight were written by Xin Su and Jie Dou.

Each experiment was organized by purpose, principle, materials and apparatus, results and discussion, designed to let students think clearly. Especially, the part of results and discussion devote students to have the ability of summarizing the experiment and thinking by themselves, then understanding the theoretical knowledge further. In the experiments, stained solutions, culture media, reagents and buffers, words and vocabularies are referenced to appendix.

It will be decided by the teachers of microbiological experiments which parts or experiments should be used in their teaching program. This text is also designed to be reference for pharmaceutic inspectors.

This is the crystallization of collective wisdom of teachers participating in the work accumulated in their teaching program for many years and possessed pharmaceutical characteristics. We hope to devote some effort to the pharmaceutical education.

It is very kind of you to correct it if there is any mistake in the text.

Editors
Aug, 2009

目　　录

Contents

第一部分　显微技术

微生物学的研究对象是肉眼无法直接分辨的微小生物，正是由于微生物的这个特性，使得显微镜成为这个学科中最为重要的工具之一。相应的，显微镜的工作原理和如何制备标本也是必须掌握的实验技能。第一部分将重点讲述前者，从最常用的明视野光学显微镜入手，讲述了光学显微镜的构造、原理及其使用方法，这其中也涵盖了暗视野光学显微镜和相差光学显微镜。在第一部分的最后又简单介绍了比光学显微镜更为有效的电子显微镜，这种显微镜主要用于观察微生物的详细内部结构和病毒的结构。

实验一　明视野光学显微镜

【实验目的】

1. 熟悉明视野光学显微镜的结构和工作原理。
2. 掌握明视野光学显微镜的使用方法，尤其是油镜的使用方法。
3. 观察几种微生物染色标本。

【实验原理】

明视野光学显微镜（以下简称“明视野显微镜”）是最常用的光学显微镜。之所以叫它“明视野”，是因为它就像你现在所看到的本页书上的文字一样，能够在较亮的背景下形成一个比较暗的图像。

1. 明视野显微镜的结构

明视野显微镜的基本结构如图 1－1 所示。

（1）镜座是显微镜的底座，是支撑显微镜全部重量的部件。

（2）镜臂用以支撑载物台和光学镜筒。需要注意的是在搬运显微镜时，需要一手握住镜臂，一手支撑底座。

（3）光源位于显微镜的基座上，由开关控制并对其亮度进行调节。显微镜不使用的情况下应将光源关掉。

（4）载物台用以放置被检标本，在载物台的中央有一小孔，可以使光线穿过。

（5）光圈和聚光器位于载物台下方。聚光器将穿过标本的光线进行汇聚，而光圈则用于调节穿过标本的光线强弱。

（6）载物台上装有一对玻片夹，镜检时用来固定标本。

（7）较大的粗调节旋钮位于镜壁两侧，用于快速升降载物台，进行粗略调焦，使用时应记住升高或降低载物台的旋钮旋转方向。

（8）较小的细调结旋钮用于载物台高度的细微调节，位于粗调节旋钮的中间。

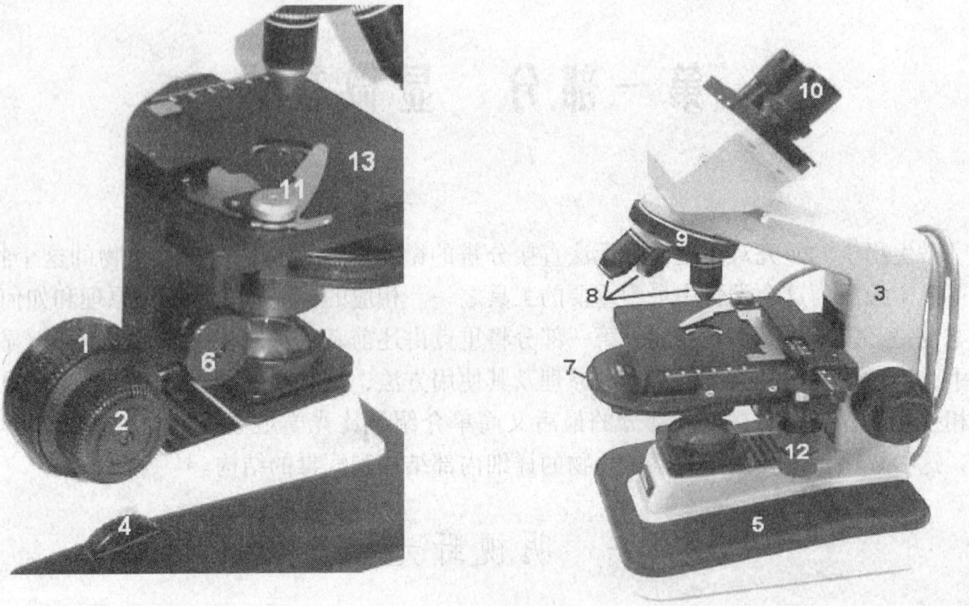

图 1－1　明视野显微镜

Fig. 1－1 Bright－field microscope

1. 粗调节旋钮；　2. 细调节旋钮；　3. 镜臂；　4. 电源开关（亮度调节旋钮）；　5. 镜座；　6. 聚光器调节旋钮；　7. 光圈；　8. 物镜；　9. 物镜转换器；　10. 目镜；　11. 玻片夹；　12. 标本推动旋钮；　13. 载物台

1. Coarse adjustment knob；　2. Fine adjustment knob；　3. Arm；　4. Power switch/brightness control　5. Base；　6. Condenser knob；　7. Iris diaphragm；　8. Objective lens；　9. Revolving nosepiece；　10. Ocular eyepiece lens；　11. Stage clips；　12. Stage motion knobs；　13. Stage

（9）光学头部包括各种镜头，与竖直方向呈一定角度。

（10）物镜转换器位于光学头部的下半部，是一个带有物镜的可旋转圆盘。

（11）显微镜有两种放大镜头，一种是目镜，一种是物镜。每一部分都对载物台上的标本起放大作用。显微镜的放大倍数由物镜和目镜的乘积决定。转换器可以将物镜进行转换，因而也就改变了显微镜的放大倍数。

转换器上共有四个物镜，分别是带有红圈的四倍物镜，10 倍的低倍镜（黄色），40 倍的高倍镜（蓝色）和 100 倍的油镜（白色）。需要注意的是这些物镜之所以用颜色进行标记是为了更容易进行辨认。在必要情况下镜头只能用擦镜纸进行清洁。

2. 显微镜的分辨率和油镜的使用

显微镜所观察到的应是放大且清晰的图像。其清晰度由分辨率来决定。分辨率（R）是指能够将非常靠近的两点进行区分的能力。分辨率越小，其分辨能力越高。下面的方程式显示了决定分辨率的主要参数：

$$R=0.61\,\frac{\lambda}{\mathrm{NA}}$$

λ 是所用的光线波长，NA 叫做数值孔径。由方程可见，所用波长越小，呈像越清晰，后面讲到的电子显微镜所具有的高分辨率就与此有关。分辨率的另一个决定因素是数值孔径。数值孔径是用来描述镜头改变光线方向的相对有效性的数学常数。数值孔径的大小如下面方程显示：

$$NA = n \sin\theta$$

其中，n 代表物镜与样本间介质的折射率，θ 则定义为穿过样本进入镜头的光锥角度的 1/2，即镜口角（图 1－2）。

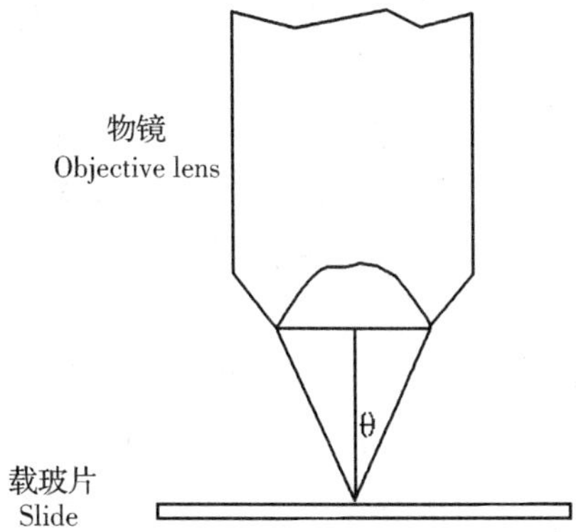

图 1－2　数值孔径

Fig. 1－2 Numerical aperture

数值孔径越大，分辨率效果越好。要提高数值孔径，可以通过提高 n 和 sinθ 来达到。空气的折射率是 1，并且由于 sinθ 不会大于 1，所以任何透镜在空气中的数值孔径都不会大于 1。要使数值孔径大于 1 的唯一可行方法就是通过使用香柏油来提高折射率，香柏油与载玻片具有相同的折射率。当用香柏油代替空气时，入射光线就不会发生反射和折射。因此，就能够得到较大的数值孔径和更好的分辨效果（图 1－3）。

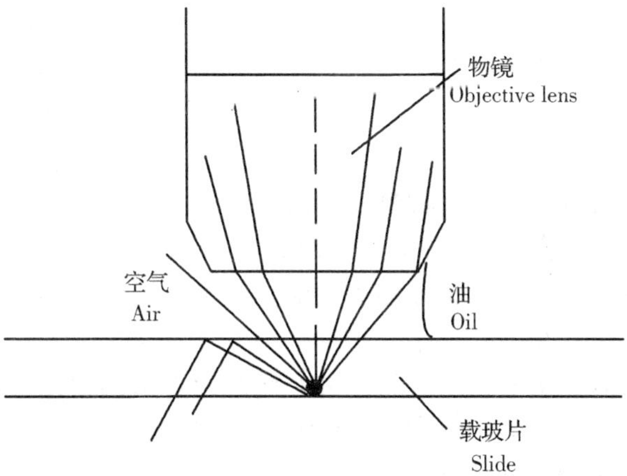

图 1－3　油镜

Fig. 1－3 The oil immersion objective lens

【实验材料】

1. 载玻片标本

（1）细菌基本形态标本　金黄色葡萄球菌、唾液链球菌、大肠埃希菌、枯草芽孢杆菌、霍乱弧菌。

（2）细菌特殊结构标本　芽孢（枯草芽孢杆菌、破伤风梭菌）、鞭毛（伤寒沙门菌）、荚膜（肺炎链球菌）。

（3）真菌基本形态标本　酿酒酵母。

2. 仪器与试剂

明视野显微镜、二重瓶（内装香柏油和二甲苯）、擦镜纸、吸水纸。

【实验步骤】

（1）一手握住镜臂，一手托住镜座，取出显微镜置于实验台上，镜座距实验台边缘约3～4cm。为减少疲劳，镜检者应姿势端正，两眼睁开。一般用左眼观察，右眼便于绘图或记录。

（2）将标本用玻片夹固定在载物台上，通过调节标本推动旋钮将涂有标本的位置置于载物台光源通过孔的正上方。

（3）适当调节光圈、聚光器和光亮度，使视野得到均匀照明。

（4）先用低倍镜（即10倍物镜）进行观察，将载物台上升至载玻片与物镜下方距离大约为5mm处，然后用粗调节旋钮缓慢下降载物台，同时从目镜中观察，直至视野中出现图像。这样做可以防止在观察过程中物镜与载玻片接触而损坏物镜。当视野中出现图像时，改用细调节螺旋将图像调至清晰。注意：细调节旋钮不可沿同一方向过度调节。

（5）转至高倍镜（即40倍物镜），此时标本仍应在视野中，只不过比原来大了4倍。使用齐焦物镜时，用细调节旋钮将物象调至清晰。如用不齐焦物镜时，每转换一次物镜都要重新调焦。从低倍镜转至高倍镜时，先将载物台上升至非常靠近载玻片的位置，然后按步骤（4）再次进行调节。真菌标本用高倍镜即可观察清楚。

（6）用油镜观察细菌的染色标本：降低载物台，在载玻片上光源透过的位置加一滴香柏油，然后将100倍物镜（即油镜）转入光路，用粗调节旋钮升高载物台，并从侧面观察，使镜头浸入油中并几乎与载玻片接触。注意：不可用力过猛，以免压碎玻片，损坏镜头。

（7）从目镜中进行观察，同时将光亮度升至最大。用粗调节旋钮缓缓下降载物台至出现物象时，改用细调节旋钮调整至物象清晰。

（8）当使用油镜时，一旦镜头接触香柏油，就不要再将高倍镜转入。否则高倍镜头会因为沾到油而需要彻底清洗。

（9）当观察结束时，用擦镜纸擦除油镜上残留的香柏油，也可用二甲苯来进行此步骤，但最后也要用擦镜纸将残留的二甲苯擦掉。显微镜镜头只能在必要时用擦镜纸清洁。

（10）将4倍物镜转入，载物台降至最低。

（11）显微镜送回存放处。

【结果与讨论】

1. 分别绘出明视野显微镜下观察到的微生物标本的典型形态结构。
2. 使用油镜时应注意哪些问题？
3. 用明视野显微镜能否观察病毒？为什么？

PART ONE Microscopic Techniques

Microbiology is the study of organisms which are too small to be seen distinctly with the unaided eye. Because of this characteristic of microbes, the microscope is an essential tool for this discipline. Thus, to understand how the microscope works and the way to prepare the specimens are of great importance. This part focuses on the former and begins with the most commonly used bright – field light microscope with its structure, principle and use. Some other types of light microscopes are also included as dark – field microscope and phase – contrast microscope. This section closes with a brief introduction of electron microscope which is more effective than light microscope to observe the detailed internal structures of large microbes or the structure of viruses.

1 Bright – Field Light Microscope

Purposes

1. To be familiar with the structure and principle of the bright – field light microscope.

2. To know how to use the microscope correctly, especially the oil immersion lens.

3. To observe several stained microorganism specimens.

Principles

The bright – field light microscope (abbreviated as "bright – field microscope") is the most commonly used type of light microscope. It is named "bright – field" as it forms a dark image against a brighter background just like the words on this page.

1. The structure of the bright – field microscope

The fundamental structure of a bright – field microscope is illustrated in figure 1 – 1.

(1) The base is the main support for the microscope.

(2) The arm supports the stage and the optical head. Note that the proper way to carry a microscope is to grasp the arm with one hand and support the base with the other hand.

(3) An electric light is mounted on the base of the microscope, and is controlled by an on – off switch which can also adjust the brightness. Please keep the light off when not in use.

(4) The stage is the flat area upon which the specimen is placed. It has a hole in the center through which light may pass.

(5) The iris diaphragm and condenser are located under the stage. The condenser focuses the light going through the specimen and the iris diaphragm is used to regulate the amount of light passing through the specimen.

(6) Mounted on the stage is a pair of stage clips. This is designed to keep the slide in place from moving while you are looking at it.

(7) A pair of large knobs, called the coarse adjustment knobs, which permit rapid raising or lowering of the stage, are located on the sides of the arm. Memorize the direction to rotate this knob to lower or raise the stage.

(8) A pair of smaller knobs, called the fine adjustment knobs, which permit smaller adjustments in the stage height, are located " inside" the coarse adjustment knobs.

(9) The optical head, the body of the microscope which contains the lenses, is bent at an angle on these microscopes.

(10) The nosepiece is a revolving plate on the lower side of the optical head which holds the objective lenses.

(11) There are two magnifying lenses in the microscope. One magnifying lens is in the ocular and one is in the objective. Each contributes to the magnification of the object on the stage. The total magnification of any set of lenses is determined by multiplying the magnification of the objective by the magnification of the ocular. The nosepiece rotates allowing the objectives to change and thus change the magnification of the microscope.

On the nosepiece, a 4 × objective or lens (red band), a 10 × low – power objective (yellow), a 40 × high – power objective (blue), and a 100 × oil immersion lens (white) may be found. Note that these lenses are color – coded, and remember which color is which for easy identification. Only lens paper should ever be used on the lenses if necessary.

2. Microscope resolution (resolving power) and the oil immersion objective

Microscope is used to provide not just a magnified image but a clear one. Resolution (R) is essential for this. Resolution is the capacity of an optical system to distinguish or separate two adjacent objects or points form one another. The smaller the resolution is, the better the revolving ability is. The following equation expresses the main determining factors in resolution:

$$R = 0.61 \frac{\lambda}{NA}$$

Lambda (λ) is the wavelength of light used to illuminate the specimen and the NA is the numerical aperture. From the equation, it indicates that a major factor in resolution is the wavelength of light used. The shorter the wavelength is, the clearer the image will be. This will help to understand the high resolution of the electron microscope later. Another factor in resolution is NA. NA is a mathematical constant that describes the relative efficiency of a lens in bending light rays. It could be expressed as:

$$NA = n \sin\theta$$

In the equation, n represents the refractive index of the medium between the objective and the specimen. θ is defined as half of the angle of the cone of light that enters a lens from a specimen (figure 1-2).

Remember that a higher NA can provide better resolution. There are two ways to increase the NA value which are higher n and $\sin\theta$. The refractive index for air is 1.00. Since $\sin\theta$ cannot be greater than 1.00, no len's working in air can have a NA higher than 1.00. The only practical way to increase the NA to above 1.00 is to increase the refractive index with immersion oil, a colorless liquid with same refractive index as the slide. When air is replaced with immersion oil, reflection and refraction of the light from the condenser to the objective will be avoided effectively. Thus, an increase in numerical aperture and resolution is achieved (figure 1-3).

Materials and Apparatus

1. Specimens on glass slides

(1) The basic morphology of bacteria - *Staphylococcus aureus*, *Streptococcus salivarius*, *Escherichia coli*, *Bacillus subtilis* and *Vibrio cholerae.*

(2) The special structure of bacteria - endospore (*Bacillus subtilis*, *Clostridium tetani*), flagella (*Salmonella typhosa*) and capsule (*Streptococcus pneumoniae*).

(3) The basic morphology of fungi - *Saccharomyces cerevisiae.*

2. Apparatus and reagents

Bright - field microscope, dual - bottle (with immersion oil and xylene), lens paper, water - adsorption paper.

Procedures

(1) Always carry the arm of the microscope with one hand and support the base

with the other hand. Place it on the desk 3 to 4cm away from the desk edge. In order to reduce fatigue, observers should posture properly and open both eyes. Usually use the left eye to observe and use the right eye to take records.

(2) Place the provided slide on the stage, and secure it firmly using the stage clips. Try to guesstimate where the specimen is located on the slide, and place it in the center of the hole allowing light through the stage.

(3) Adjust the iris diaphragm, condenser, and brightness of light carefully to get suitable illumination.

(4) With the low - power objective (10 × objective) in position, raise the stage until the tip of the objective is within 5mm from the slide. Then lower the stage slowly using the coarse adjustment knobs while looking at the microscope until the object comes into view. This can avoid striking the objective to the slide. Once the object is in view, switch over to the fine adjustment knob to focus the desired image. Note: Do not overuse the fine adjustment knob in one direction.

(5) Rotate the high - power objective (40 × objective) in position. Your specimen should still be seen in the field of vision, but 4 times larger now. When using the microscope that has equal focus for the objectives, clarify the image with your fine adjustment knobs. For the microscope that does not have this feature, when switch over to the high - power objective (or any other objective), raise the stage until the slide almost touch the lens and then do the same manipulation as step (4) Fungi specimen can be observed clearly with the high - power objective.

(6) Use the oil immersion lens to examine the stained bacteria samples provided. Lower the stage and place a single drop of immersion oil on the slide right over where the light is coming through the stage. Then rotate the 100 × objective (oil immersion lens) into position. Raise the stage with the coarse adjustment knob while looking at the objective from the side until the lens immerses in oil and nearly touches the slide. Note: Do this gently to avoid breaking the slide and the objective.

(7) Now look through the oculars, increasing your light for maximum illumination. Lower the stage slowly with the coarse adjustment knob. Once the object is in view, use the fine adjustment knob to focus clearly.

(8) Once you have gone into immersion oil, do not go back to the 40 × objective. The objective will get oil on it, and you will have to really clean it to get the oil off.

(9) After you are finished with the microscope, use the lens paper to wipe the oil from the 100 × objective lens. Xylene could be used to clean the oil if necessary, but remember to remove xylene with the lens paper as well. Clean all the microscope's lenses only with lens paper if necessary.

(10) Rotate the 4 × objective lens into place. Use the coarse adjustment knobs to lower the stage to the bottom.

(11) Return the microscope to the appropriate storage area.

Results and Discussion

1. Draw the typical morphology of microorganisms you observed through the bright – field microscopes.
2. What should be paid attention to when you use the oil – immersion lens?
3. Can you observe virus with the bright – field microscope? Why?

实验二　暗视野光学显微镜

【实验目的】

1. 了解暗视野光学显微镜的基本原理。
2. 正确使用暗视野光学显微镜。

【实验原理】

暗视野光学显微镜（以下简称“暗视野显微镜”）与明视野显微镜相比，除聚光器外其他结构都相同。如果将明视野显微镜的聚光器镜头下方安装一个暗视野光阑，就可以将其改造成一个暗视野显微镜（图 2－1）。

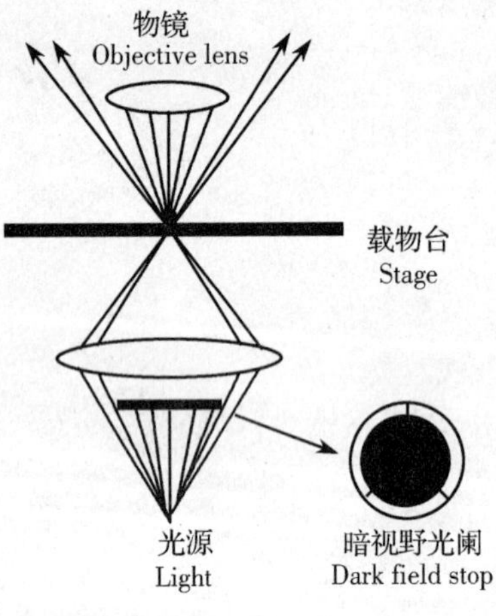

图 2－1　暗视野显微镜光路

Fig. 2－1 Light pathway of a dark－field microscope

暗视野光阑使得只有由标本反射或折射的光线才能进入到物镜中。当光线经过暗视野聚光器时，形成了一个中空的光锥照射在标本上并成像。此时标本周围的视野呈黑色，而标本本身被照亮。由于在黑色背景下观察亮的物体总是比在相反条件下所看到的图像更清晰，因此暗视野显微镜更适于观察未染色或难于染色的微生物标本，同时也适用于观察微生物的外部轮廓以及判断细胞的运动性。暗视野显微镜的分辨能力也比明视野显微镜要高得多。

【实验材料】

1. 菌种

酿酒酵母培养物，枯草芽孢杆菌培养物。

2. 仪器与试剂

暗视野显微镜，载玻片，盖玻片，无菌水，双重瓶，擦镜纸，吸水纸。

【实验方法】

1. 直接在暗视野聚光器上滴加一滴香柏油。
2. 制备酵母细胞水浸片。然后将标本片放于载物台上，使标本处于光源上方。
3. 小心上升聚光器，直至油滴接触到载玻片。
4. 先用低倍镜进行观察，然后转至高倍镜，用粗调节旋钮和细调节旋钮将图像调至清晰。
5. 按照相同步骤观察枯草芽孢杆菌标本，最后用油镜进行观察。

【结果与讨论】

1. 分别记录酿酒酵母与枯草芽孢杆菌观察结果。
2. 与明视野显微镜相比，暗视野显微镜更适于观察活细胞微生物。为什么？

2 Dark – Field Light Microscope

Purposes

1. To understand the principles behind the dark – field light microscope.
2. To know how to use the dark – field light microscope correctly.

Principle

The dark – field light microscope (abbreviated as "dark – field microscope") has the same structure as the bright – field microscope except the condenser. A bright – field microscope can be converted to a dark – field microscope by placing a dark – field stop underneath the condenser lens (figure 2 – 1).

The stop blocks all light from entering the objective lens except the light that has been reflected or refracted by the specimen. A hollow cone of light is now focused on the specimen and forms an image. The field surrounding a specimen appears black, while the object itself is brightly illuminated. Since light objects against a dark background are seen more clearly by the eye than the reverse, dark – field microscopy is useful in observing both unstained living microbes and microbes that are difficult to stain. It can outline the organism's shape and recognize the cellular motility. The resolving ability of the dark – field microscopy is much greater than that of the bright – field microscopy.

Materials and Apparatus

1. Cultures

Cultures of *Saccharomyces cerevisiae* and *Bacillus subtilis.*

2. Apparatus and reagents

Dark – field microscope, slides, cover slips, sterile water, dual – bottle, lens paper, water – adsorption paper.

Procedures

1. Place a drop of immersion oil directly on the dark – field condenser lens.

2. Prepare a wet mount of yeast cells. Then put it on the stage and make sure that the specimen is just over the light opening.

3. Raise the condenser carefully until the oil just touches the slide.

4. Observe with the low - power objective first and then switch to the high - power objective. Focus with the coarse and fine adjustment knobs.

5. Do the same with *B. subtilis* specimen. And then use the oil - immersion lens to observe.

Results and Discussion

1. Record the observation result of the two specimens respectively.

2. Compared with bright - field microscope, the dark - field microscope is more - suitable for observing living microorganism. Why?

实验三 相差光学显微镜

【实验原理】

了解相差光学显微镜的基本原理并学会如何正确使用。

【实验原理】

使用明视野显微镜并不能够清楚观察未染色的活细胞及其内部结构，这是因为细胞与水环境或细胞内部结构之间的对比度太小。不过细胞的不同结构之间确实存在密度上的细微差别，从而能够将通过其中的光线有所改变。相差光学显微镜（以下简称“相差显微镜”）可以很容易的将这种细微的改变转化成为可检测出的光强变化。因此这种显微镜非常适用于观察活细胞。

与普通光学显微镜相比，相差显微镜有另外两个部件：一个是位于物镜后聚焦面的相板；另外一个是与相板相匹配的位于聚光器上的环形光阑，它是一个黑色圆盘，上面带有一个透明圈。相板上有一环状区域，经过其中的光线不会发生改变，而经过环状区域以外的其他区域的光线则会被延迟 1/4 波长（λ）。

带有环状光阑的聚光器可以产生中空的光锥。当光锥通过细胞的时候，部分光线会由于标本中各结构密度与折射率的不同发生偏斜，并被推迟 1/4λ，由这些光线汇聚形成了备检物体的一个图像。未发生偏斜的直射光照射到位于相板上的环形区域，而衍射光则会错过这一区域而从相板上的其他区域穿过。由于相板可以使通过的衍射光再被推迟 1/4λ，直射光与衍射光之间就会存在 1/2 的相位差，当两者汇聚在一起时，这部分就会相互抵消而呈现出较暗的图像（图 3-1）。

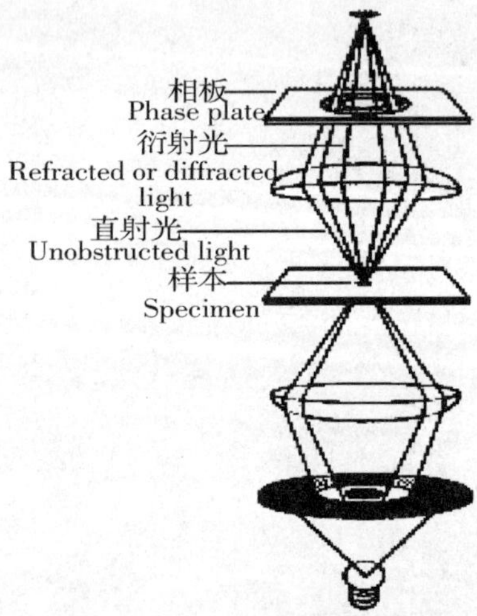

图 3-1　相差显微镜光路

Fig. 3-1 Light pathway of a phase-contrast microscope

【实验材料】

1. 菌种

枯草芽孢杆菌培养物，酿酒酵母培养物。

2. 仪器与试剂

相差显微镜，载玻片，盖玻片，无菌水，双重瓶，擦镜纸，吸水纸。

【实验方法】

1. 制备枯草芽孢杆菌水浸片。

2. 将标本置于载物台上。

3. 先转至10倍物镜，同时将环形光阑也调至10倍位置，用粗调节旋钮和细调节旋钮将物像调至清晰。

4. 将目镜取下，换上合轴调整望远镜，调整至亮环与暗环都清晰，固定合轴调整望远镜。

5. 调节聚光器至两环相互重叠（图3－2）。

6. 将目镜重新安上，用细调节旋钮将图像调节清晰并观察。

7. 按同样的方法分别用40倍物镜和油镜观察，同时要将环形光阑也调至相应位置。记录观察结果。

8. 按照相同的步骤在40倍物镜下观察酿酒酵母。

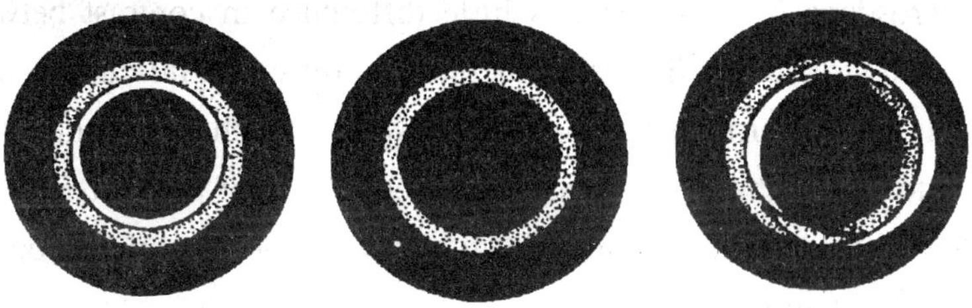

A　　　　B　　　　C

图3－2　相差显微镜的调节

Fig. 3－2 Adjust the phase－contrast microscope condenser

A. 调节亮环和暗环清晰；　B. 两环重合；　C. 两环不重合

A. Adjust the dark ring and the bright ring clearly　B. The dark ring and the bright ring superimposition　C. The dark ring and bright ring do not coincide

【结果与讨论】

1. 分别记录两种微生物样本的观察结果。

2. 为什么可以用相差显微镜观察活细胞的细微结构？

3 Phase – Contrast Light Microscope

Purposes

To understand the basic principles behind the phase – contrast light microscope and know how to use the phase – contrast microscope correctly.

Principles

Unstained living cells and their internal components are not clearly visible in the bright – field microscope because there is little difference in contrast between cell and water or the internal structures. But cell structures do differ slightly in density which can alter the light that passes through them in subtle ways. A phase – contrast light microscope (abbreviated as "phase – contrast microscope") can converts the slight differences into easily detected variations in light intensity. Thus this microscopy is an excellent way to observe living cells.

Compared with the bright – field microscope, phase – contrast microscope has two additional components: a "phase plate" located at the back focal plane of the objective lens and a matching "annular diaphragm" in the condenser consisting of a clear ring on a black field. There is a ring – shaped area which is on the phase plate. Light can be retarded exactly 1/4 wavelength by the area except the ring on the phase plate.

The condenser with annular diaphragm can produce a hollow cone of light. As this cone passes through a cell, some light rays are bent due to variations in density and refractive index within the specimen and are retarded by about 1/4 wavelength. The deviated light is focused to form an image of the object. Undeviated light rays strike a phase ring in the phase plate while the deviated rays miss the ring and pass through the rest of the plate. Because the phase plate is constructed in such a way that the deviated light passing through it is advanced by 1/4 wavelength, the deviated and undeviated waves are now about 1/2 wavelength out of phase and will cancel each other when they come together to form an image (Fig. 3 – 1).

Materials and Apparatus

1. Cultures

Cultures of *Bacillus subtilis* and *Saccharomyces cerevisiae.*

2. Apparatus and reagents

Phase – contrast microscope, slides, cover slips, sterile water, dual – bottle, lens paper, water – adsorption paper.

Procedures

1. Make a wet – mount of *B. subtilis.*
2. Place the slide on the stage.
3. Rotate the 10 × objective into place as well as the annular diaphragm that corresponds to the 10 × objective. Adjust the coarse and fine adjustment knobs to focus the image.
4. Replace the ocular lens with a phase centering telescope. Adjust the telescope until the dark ring and bright ring are clear (Fig. 3 – 2A). Fix the telescope.
5. Adjust the condenser to make the two rings coincide (Fig. 3 – 2B).
6. Put the ocular lens back. Use the fine adjustment knob to clarify the image and observe.
7. Do the same with the 40 × objective and the oil immersion lens. Remember to adjust the corresponding diaphragm in position. Record the results.
8. Do the same with *S. cerevisiae* under 10 and 40 × objectives.

Results and Discussions

1. Record the observation results of the two specimens respectively.
2. Why can the phase – contrast microscope be used to observe the microstructure of living cells?

实验四　电子显微镜

光学显微镜的最高分辨率大约为 0.2μm，因此只能用来观察细菌的一般形态特点，对于细菌内部结构或是病毒的构造则无能为力。电子显微镜就能够克服这些不足，常用的电子显微镜有两种：透射电子显微镜（TEM）和扫描电子显微镜（SEM）。本实验着重介绍前者。

【实验目的】

1. 掌握透射电子显微镜的原理与结构。
2. 学习如何制备电镜标本。

【实验原理】

透射电子显微镜结构复杂，但其基本操作原理与光学显微镜相似，只不过电子显微镜以电子束代替光线，以磁场代替透镜（图 4-1）。光学显微镜在观察标本时主要受到使用光波波长的限制。透射电镜以电子束作为“光源”，这种极短的波长使得其分辨率能够比光学显微镜提高数千倍。

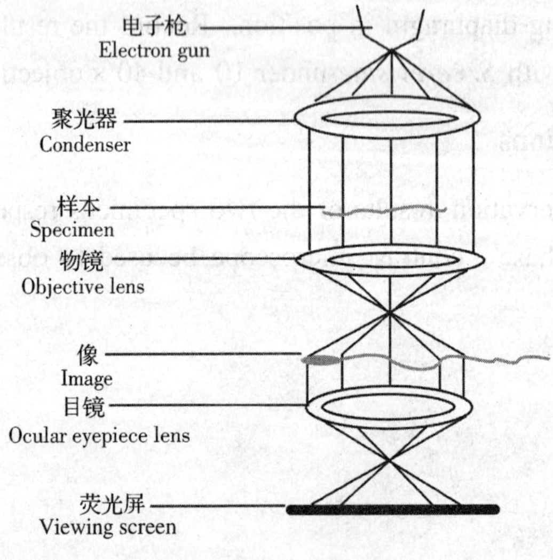

图 4-1　透射电镜示意图
Fig. 4-1 TEM diagram

在真空条件下，由电子枪中加热的钨丝发射出电子束。与光学显微镜中用玻璃透镜汇聚光线不同的是，透射电镜用电磁场将电子束进行汇聚。当电子束穿过标本时，一部分电子会发生散射，密度较大的部位会散射更多电子，呈像就会较暗，而电子穿透过的区域则比较明亮。电子束经过磁场透镜汇聚后就会在荧光屏上呈现出被放大的图像，可直接用于研究或利用相机拍摄下来。

透射电镜用于观察细胞的内部结构或者病毒构造。由于电子不能穿过厚的物体，因此被检样本必须制备成超薄切片，并用金属染色以增强其图像对比度。

标本的制备方法由所用电镜类型及标本性质决定。常用方法之一就是负染色，利用金属原子作为电子染料包裹在未染色的生物学样品外来成像。重金属可以使拍摄结果中的背景变暗，而样品呈现出光亮的效果。本实验将介绍如何制备轮状病毒负染色标本。

【实验材料】

1. 试剂

聚乙二醇，2%磷钨酸（pH6.5～8.0）水溶液，0.5%聚乙烯醇缩甲醛（Formvar），轮状病毒（*Rotavirus*）悬液。

2. 仪器

铜网若干，滤纸，普通光学显微镜，透射电镜。

【实验方法】

1. 取经聚乙二醇浓缩的病毒悬液，滴在有Formvar膜的铜网上。

2. 以滤纸吸收过多标本，然后趁铜网未干燥时滴上磷钨酸溶液，染色5分钟。

3. 将多余染液以滤纸吸干，待样品干燥后，先置低倍光学显微镜下检查，挑选制片效果较好的铜网置透射电镜观察。

4. 在电镜荧光屏上，可见到灰暗背景下类似车轮状的明亮的病毒颗粒（图4－2）。

图4－2　轮状病毒结构模式图

Fig. 4－2 The model of *Rotavirus*

【结果与讨论】

绘制轮状病毒的负染色图像。

4 Electron Microscope

The best light microscope has a resolution limit of about 0. 2μm. Only the bacterial general morphological features are visible in the light microscope. The internal structures of large microbes or the viral structure cannot be studied intensively. Electron microscopes can overcome these shortages of light microscopes. Two commonly used electron microscopes are the transmission electron microscope (TEM) and the scanning electron microscope (SEM). This experiment focuses on the former.

Purposes

1. To understand the principle and structure of the transmission electron microscope.

2. To know how to prepare the specimen for electron microscope.

Principles

A modern TEM is complex, but the basic principles behind its operation are almost the same as the light microscope except that TEM uses electrons instead of light and magnetic lens instead of the glass lens (Figure 4 – 1). What you can see with a light microscope is limited by the wavelength of light. TEM uses electrons as "light source" and its much lower wavelength makes it possible to get a resolution a thousand times better than with a light microscope.

A heated tungsten filament in the electron gun emits the electrons that travel through vacuum in the column of the microscope. Instead of glass lenses focusing the light in the light microscope, the TEM uses electromagnetic lenses to focus the electrons into a very thin beam. The electron beam then travels through the specimen. Some of the electrons are scattered when passing through the specimen. A denser part scatters more electrons and shows darker in the image while the electron – transparent regions are brighter. The beam is focused by magnetic lenses to form an enlarged, visible image of the specimen on a fluorescent screen. The image can be studied directly by the operator or photographed with a camera.

TEMs are used for viewing the detailed structure of cells and viruses. Because electrons can not readily penetrate thick preparations, the specimen must be sectioned into extremely thin slices and stained with metals that will increase image contrast.

Specimen preparation methods vary according to the type of electron microscope used and the nature of the sample. One very useful technique is negative staining. It uses metal atoms as electron dye to wrap the unstained biological sample to form an image. Heavy metals render the background dark, whereas the specimen appears bright in photographs. In this exercise, the specimen preparation of *Rotavirus* for negative staining is introduced.

Materials and Apparatus

1. Reagents

PEG, 2% PTA (pH6.5 ~ 8.0), 0.5% Formvar, *Rotavirus* suspension.

2. Apparatus

Copper nets, filter paper, bright - field light microscope, the transmission electron microscope.

Procedures

1. Use PEG to concentrate the suspension of *Rotavirus* and drip on copper nets with a Formvar film.

2. Remove the excessive suspension with filter papers and then stain with PTA for 5min before the copper nets dry.

3. Remove the surplus PTA with filter papers. Waiting to dry, observe with light microscope to choose an appropriate copper net with better preparation condition. Then observe it with the TEM.

On the fluorescent screen of the TEM, virus particles appear bright wheels against a dark background (Figure 4 - 2).

Results and Discussions

Draw the negative stained *Rotavirus*.

第二部分 微生物形态学

实验五 细菌单染色法

【实验目的】

1. 熟悉细菌单染色标本的制备过程。
2. 用油镜观察细菌的形态结构。

【实验原理】

由于细菌微小，加之它与周围的水环境光学性质相近，从而用一般光学显微镜不易看到它。通常用染色的方法增加反差，有助于细菌标本的观察。染料有带阴离子发色团的酸性染料和带有阳离子发色团的碱性染料。由于在一般生理条件下（pH7.4 左右）细菌菌体都带负电荷，从而更容易与碱性染料相结合。常用的碱性染料有结晶紫、美蓝和碱性复红。

染色法又分单染色和复染色两大类。单染色法即仅用一种染料着色，所有的细菌均被染成一种颜色，可用来观察细菌的形态和排列方式，但无鉴别细菌的作用。复染色法又称鉴别染色法，通常用二种或二种以上的染料着色，由于不同种类的细菌或同种细菌的不同组成结构对染料有不同的反应性而被染成不同的颜色，从而有鉴别细菌的作用。

【实验材料】

1. 细菌

金黄色葡萄球菌斜面培养物 1 支，大肠埃希菌斜面培养物 1 支。

2. 试剂

吕氏美蓝染色液，复红染色液，香柏油，生理盐水，二甲苯。

3. 仪器

载玻片，吸水纸，接种环，酒精灯，擦镜纸，显微镜。

【实验方法】

1. 涂片

（1）取载玻片一张，拭净。

（2）接种环火焰灭菌，取生理盐水一滴，放在载玻片中央（如被检材料是液体，可不加生理盐水）。

（3）左手斜持菌种管，右手持接种环，经火焰灭菌后，用右手小指拔开菌种管棉塞，管口通过火焰，将接种环插入管中取菌少许（切不可多，更不可将培养基刮下）。

（4）管口再通过火焰，塞好棉塞。

（5）将接种环上的细菌加入载玻片上之水滴内，磨匀，涂成直径约 1cm 大小的薄菌膜。

（6）接种环经火焰灭菌。

2. 干燥

涂片置于空气中，使其自然干燥。

3. 固定

干燥后将涂片在火焰上缓缓通过 3 次，此为“固定”。目的是使细菌粘于载玻片上，染色和水冲时不易脱落；且细菌为蛋白质，被热凝固可保持完整形态。

4. 染色

（1）在固定后的标本上加吕氏美蓝染液（或复红染液）以覆满标本为度，染 1 ~2 分钟。

（2）用细流水自载玻片一端徐徐冲洗。

（3）待其自然干燥或用吸水纸轻轻吸干。

细菌单染色法见图 5 －1。

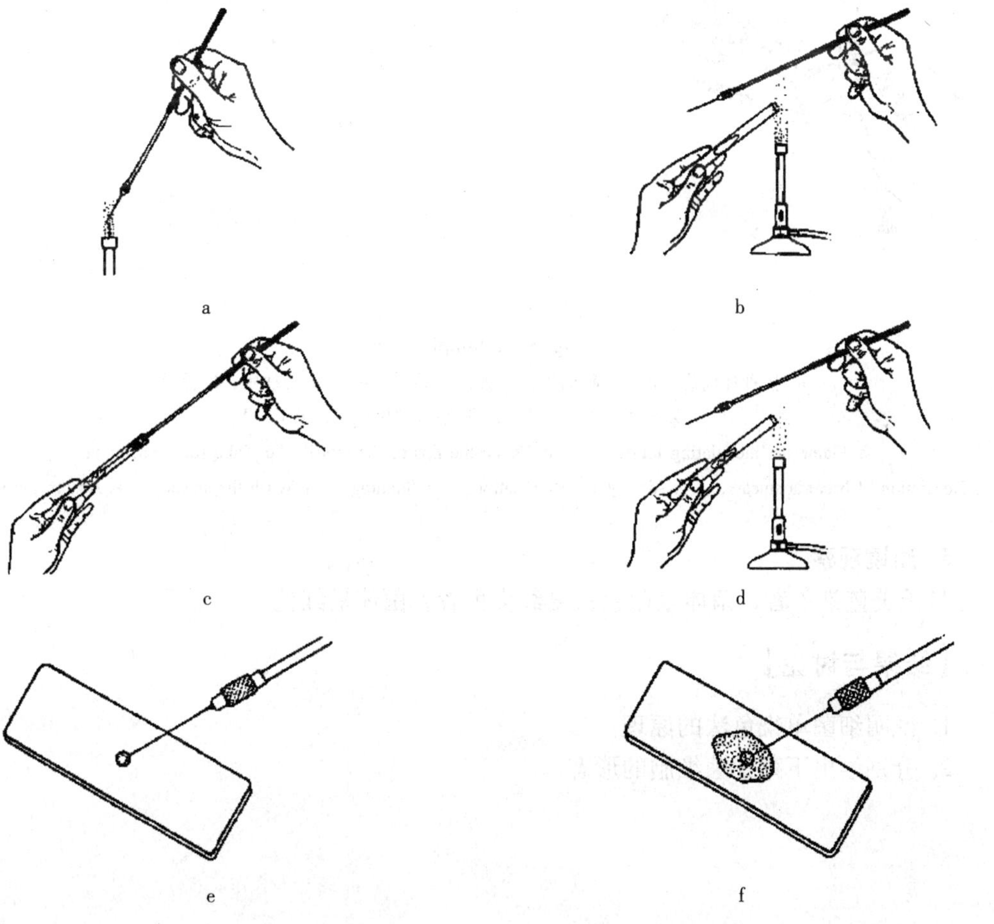

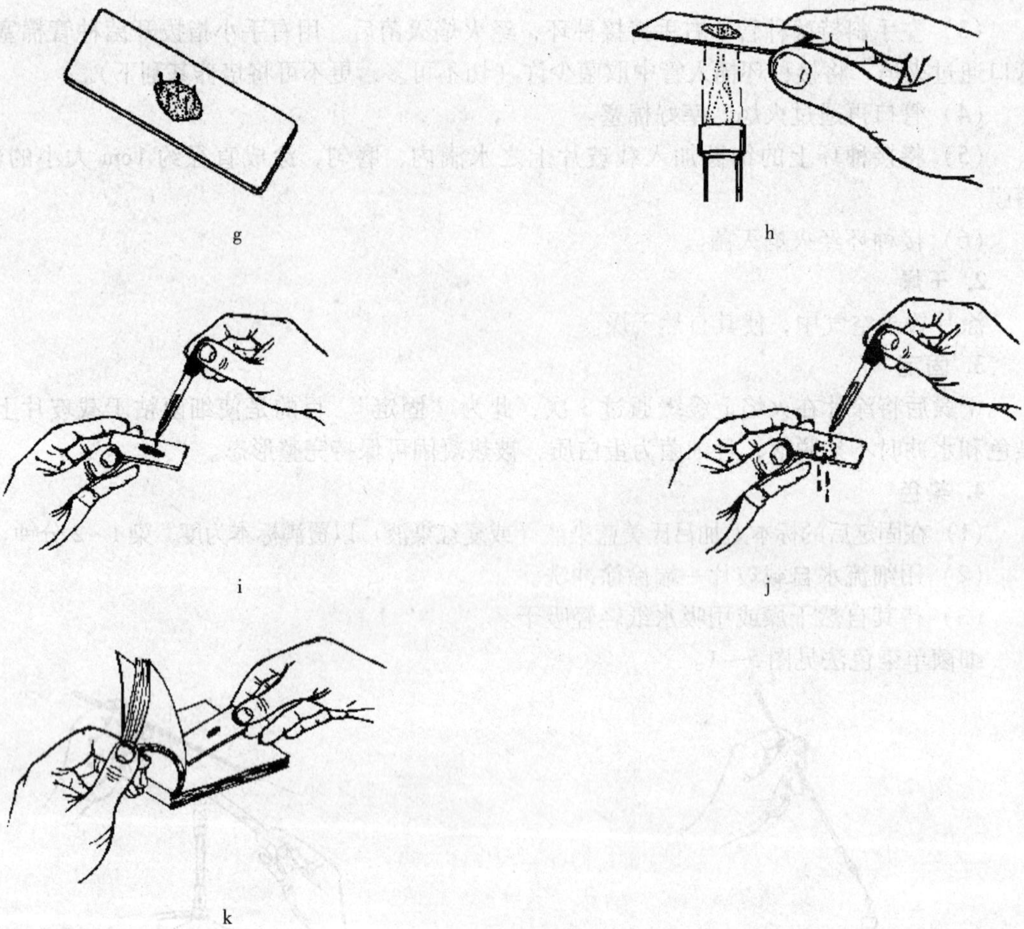

图 5－1　细菌单染色法

Fig. 5－1 Simple stain

a. 接种环灼烧；b、d. 斜面管口过火；c. 取菌；e. 取生理盐水；f. 涂片；
g. 干燥；h. 固定；i. 染色；j. 冲洗；k. 吸干

a. Flame the inoculating loop；　b、d. Flame the lips of the tube；　c. Take some bacteria；
f. Preparation of bacteria smears；　g. Air dry ；　h. Fixation；　i. Staining；　j. Wash the smear；　k. Absorb the water

5. 油镜观察

吕氏美蓝染色者，菌体呈蓝色；复红染色者，菌体呈红色。

【结果与讨论】

1. 说明细菌单染色法的原理。
2. 分别绘出下列细菌细胞的形态。

金黄色葡萄球菌

大肠埃希菌

枯草芽孢杆菌

PART TWO Microbial Morphology

5 Simple Stain

Purpose

1. To be familiar with the simple stain of bacteria.
2. To observe the morphology of bacteria with microscope.

Principle

Microorganism is very small, measured in μm or nm. In general, bacteria couldn't be observed with bright – field microscope due its being small and similar optical properties with water around it. After simple staining, the bacterial smear is stained with dye and can be visualized with microscope. Basic stains with a positively charged chromogen are preferred because bacterial cells carry negative charge that strongly attracts and binds to the cationic chromogen at pH 7.4. The most commonly used basic stains are crystal violet, methylene blue, and carbol fuchsin.

Sample stain is divided into simple staining and compound staining. Only one dye is used in simple staining and bacteria are dyed to be the same color. The purpose of simple staining is to elucidate the morphology and arrangement of bacterial cells but not to distinguish different bacteria. Compound staining is also named distinguish staining and two kinds of dyes are used at least. According to the reacting activity between bacteria and dyes, bacteria can be distinguished and divided.

Materials and Apparatus

1. Cultures

Nutrient agar slant cultures of *Staphylococcus aureus*, *Escherichia coli* and *Bacillus subtilis.*

2. Reagents

Methylene blue, crystal violet, fuchsin, physiological saline, cedar wood oil, xylene.

3. Apparatus

Glass slides, water - adsorption paper, inoculating loop, lamp, lens paper, microscope.

Procedures

1. Preparation of bacterial smears

(1) Clean a glass slide and place it on the staining tray.

(2) Flame the inoculating loop until the entire wire is red. Drop a blob of physiological water on the center of glass slide unless the sample is liquid media.

(3) Hold the slant cultures with left hand. Flame the inoculating loop again. Uncap the tubes with free fingers of the hand holding the sterile inoculating loop and carefully flame the neck of the tube. Take some bacteria from slant.

(4) Flame the neck of the tube and recap.

(5) Place the bacterial culture on the loop into physiological water, stirred and smear it to form a bacterial membrane of approximate 1cm diameter.

(6) Flame the inoculating loop.

2. Air dry

Let the smear to air dry.

3. Fixation

While holding the glass slide at one end, pass the smear over the flame of the lamp three times. Fixation of bacteria on the surface of the glass slide is necessary for sample dying, washing, and keeping the arrangement of bacteria while cell protein being heat - denatured.

4. Staining

(1) Stain the smear with methylene blue or crystal violet or fuchsin for 1 to 2 min.

(2) Wash the smear with tap water from one side of the glass slide to remove excess stain.

(3) Air dry or adsorb water with paper but do not wipe the slide.

5. Examine all stained slides under oil immersion.

Results and Discussion

1. What's the principle of simple staining techniques?
2. Draw a representative field for each organism.

Staphylococcus aureus

Escherichia coli

Bacillus subtilis

实验六 细菌革兰染色法

【实验目的】

1. 熟悉细菌的革兰染色法。
2. 用油镜观察革兰阳性细菌和革兰阴性细菌的形态结构。

【实验原理】

最常用的细菌复染色法是革兰染色法（Gram's stain）。可分为结晶紫初染、路哥碘液媒染、酒精脱色和复红复染等步骤。采用革兰染色法可把细菌分成两大类，它是细菌分类和鉴定的基础。凡能使第一种染料结晶紫保留蓝紫色的细菌叫做革兰阳性（G^+）细菌，凡被酒精脱色后染上对比颜料沙黄或稀释复红而呈红色的细菌叫做革兰阴性（G^-）细菌。

一般认为，革兰染色法与下列诸因素有关：① 革兰阳性细菌等电点低（pI＝2～3），而革兰阴性细菌等电点高（pI＝4～5），因此在一般生理条件下（pH7.4 左右），革兰阳性细菌所带的负电荷要比阴性菌多得多，从而与碱性染料结晶紫结合牢固。② 革兰阴性细菌细胞壁有外膜结构，含有较多的脂质成分，对酒精作用敏感。脂质被酒精溶解，造成细胞壁破损，结晶紫－碘复合物容易被抽提出来而脱色。③ 革兰阳性细菌细胞壁脂质含量低，对酒精作用不敏感。且革兰阳性细菌细胞壁含有多层致密（交联度大）的肽聚糖层以及带有大量负电荷的磷壁酸，结晶紫－碘复合物与细胞壁结合紧密，染料不易被酒精抽提出来，仍保留结晶紫的蓝紫色。

【实验材料】

1. 细菌

金黄色葡萄球菌斜面培养物 1 支，大肠埃希菌斜面培养物 1 支。

2. 试剂

结晶紫、95% 乙醇溶液、路哥碘液、稀释复红各 1 瓶，香柏油，二甲苯，生理盐水。

3. 仪器

载玻片，吸水纸，接种环，酒精灯，擦镜纸，显微镜。

【实验方法】

1. 涂片、干燥、固定

同单染色法。

2. 初染

加结晶紫染液于标本上，使其覆满标本，染 1～2 分钟，细水冲洗。

3. 媒染

加路哥碘溶液经 1 分钟，细水冲洗。

4. 脱色

加 95% 乙醇溶液于载玻片上，脱色约 30 秒，倾去酒精，脱色 1 次，细水冲洗。

5. 复染

加稀释复红染液复染约 1 分钟，水洗，待其自然干燥或用吸水纸轻轻吸干。

6. 镜检

油镜观察。

细菌革兰染色法见图 6－1。

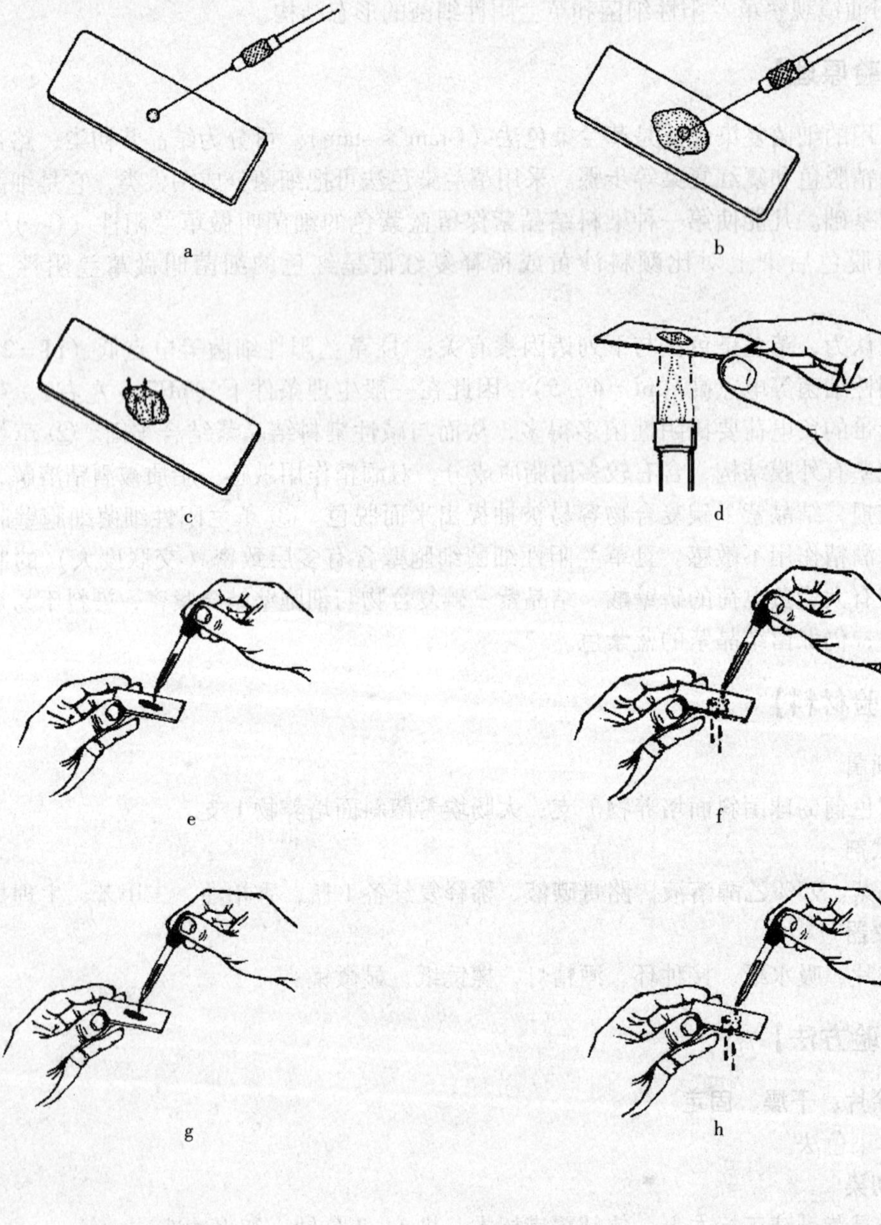

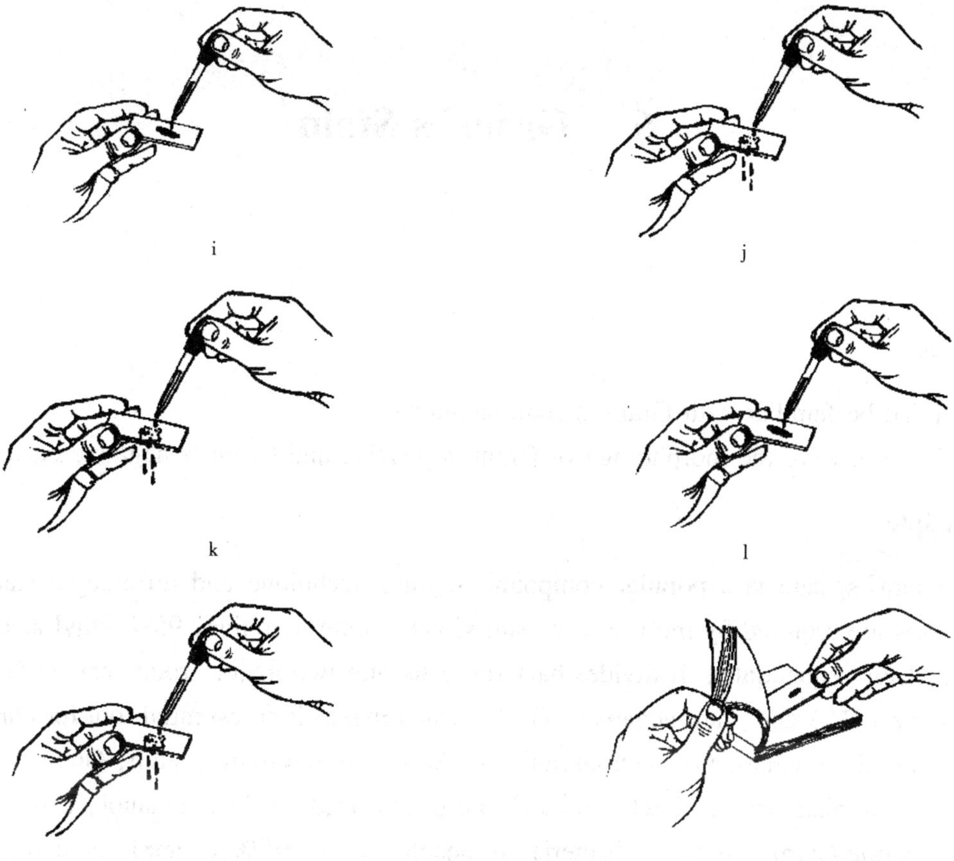

图 6 – 1　细菌革兰染色法

Fig. 6 – 1 Gram's stain

a. 取生理盐水；　b. 涂片；　c. 干燥；　d. 固定；　e. 初染；　f、h、j、k、m. 冲洗；　g. 媒染；　i. 脱色；　l. 复染；　n. 吸干

a. Get some physiological saline；　b. Preparation of bacteria smears；　c. Air dry；　d. Fixation；　e. Primary stain；　f、h、j、k、m. Wash the smear；　g. Mordant；　i. Decolorizing

【结果与讨论】

1. 说明细菌革兰染色的原理。
2. 分别绘出下列革兰阳性细菌或革兰阴性细菌的形态。

金黄色葡萄球菌　　　大肠埃希菌　　　混合涂片

6 Gram's Stain

Purpose

1. To be familiar with Gram's stain techniques.
2. To observe the morphology of Gram's positive and Gram's negative bacteria.

Principle

Gram's stain is a popular compound staining technique and differential staining processes are required by the use of crystal violet, Gram's iodine, 95% ethyl alcohol, and fuchsin sequentially. It divides bacterial cells into two major groups named Gram's positive (G^+) and gram negative (G^-), which makes it an essential tool for classification and differentiation of microorganisms. According to Gram's stain, Gram's positive bacteria bind crystal violet, couldn't be decolorized by 95% ethanol, and appear violet, while Gram's negative bacteria are decolorized by 95% ethanol, bind fuchsin, and appear red.

Generally, the principle of Gram's stain are as follows. ①The isoelectric point of Gram's positive bacteria is lower ($pI = 2 \sim 3$) than that of gram negative bacteria ($pI = 4 \sim 5$). So it contains much more negative charge and binds to crystal violet tightly. ②The cell wall of Gram's negative bacteria is composed of outer membrane containing much lipid composition, which is sensitive to ethyl alcohol. When decolorizing with 95% ethanol, the complex of crystal violet and iodine is extracted from cell wall. ③ The cell wall of Gram's positive bacteria contains less lipid composition that is not sensitive to ethanol decolorizing. In addition, it contains multiplayer of high percentage of peptidoglycan and teichoic acid with large amount of negative charges. The complex of crystal violet and iodine binds to cell wall tightly and can not be extracted by 95% ethanol. So it appears purple.

Materials and Apparatus

1. Cultures

24 h nutrient agar slant cultures of *Staphylococcus aureus* and *Escherichia coli.*

2. Reagents

Crystal violet, 95% ethanol, Lugol's iodine, fuchsin, physiological saline, cedar wood oil, xylene.

3. Apparatus

Glass slides, water-adsorption paper, inoculating loop, lamp, lens paper, microscope.

Procedures

1. Preparation of bacterial smears, air dry, and fixation refer to simple staining techniques.

2. Primary stain: Stain the smear with crystal violet for 1 to 2 min, and wash the smear with tap water from one side of the glass slide to remove excess stain.

3. Mordant: Add Lugol's iodine to the smear for 1 min, and wash the smear with tap water from one side of the glass slide to remove excess iodine.

4. Decolorizing: Add 95% ethanol to the smear, decolorize the smear for 30s once, and wash the smear with tap water from one side of the glass slide to remove excess ethanol.

5. Counterstain: Stain the smear with fuchsin for 1 min, wash the smear with tap water from one side of the glass slide to remove excess fuchsin. Air dry or adsorb water with paper but do not wipe the slide.

6. Examine all stained slides under oil immersion.

Results and Discussion

1. What's the principle of Gram's stain?

2. Draw a representative field for Gram positive and Gram negative bacteria.

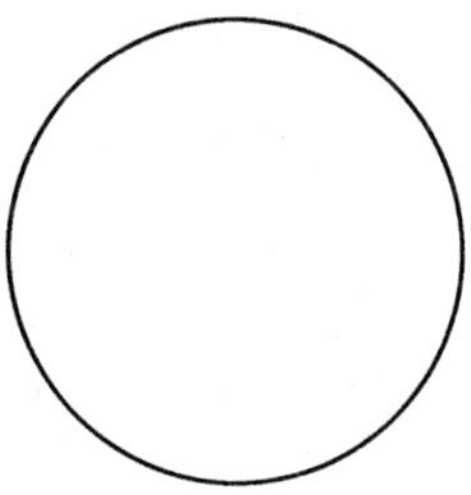

Staphylococcus aureus

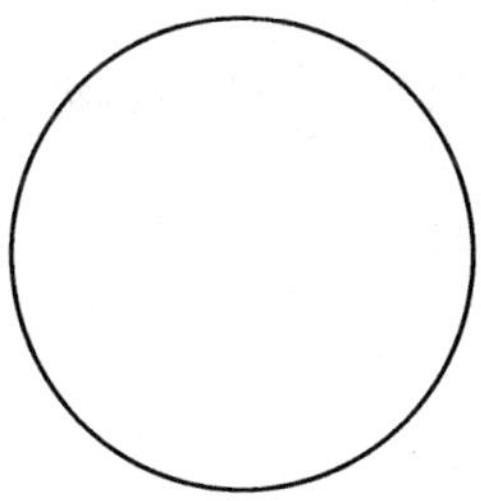

Escherichia coli

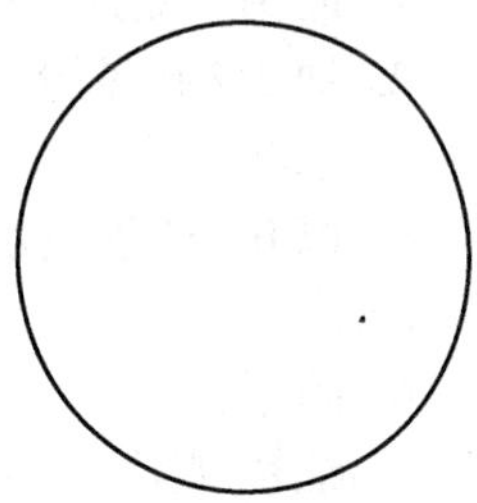

Mixtures

实验七 芽孢染色法

【实验目的】

1. 熟悉细菌芽孢的染色方法。
2. 用油镜观察细菌芽孢的形态结构。

【实验原理】

细菌芽孢具有致密的多层壁膜结构，通透性低，难以染色，故需采用加温染色法。由于细菌菌体和芽孢对染料的亲和力不同，采用不同的染料染色，可使芽孢和菌体呈不同的颜色。芽孢染色包括孔雀绿初染和沙黄复染。当先用弱碱性染料孔雀绿在加热条件下进行染色时，染料同时进入菌体和芽孢。经水洗后，菌体脱色，而芽孢由于其通透性低，染料不能被洗出。当用沙黄染液复染后，镜检可见芽孢呈绿色，菌体呈淡红色。

【实验材料】

1. 细菌

枯草芽孢杆菌 48 小时琼脂斜面培养物 1 支。

2. 试剂

5% 孔雀绿水溶液，0. 5% 沙黄水溶液，生理盐水，香柏油，二甲苯。

3. 其他

载玻片，吸水纸，接种环，酒精灯，擦镜纸，显微镜。

【实验方法】

1. 取菌、涂片、固定（同单染法）。
2. 加 5% 孔雀绿水溶液，加温染色 1 分钟，冷却。
3. 水洗。
4. 以 0. 5% 沙黄水溶液复染 30 秒。
5. 水洗。
6. 晾干。
7. 油镜观察，芽孢呈绿色，菌体呈淡红色。

芽孢染色法见图 7 – 1。

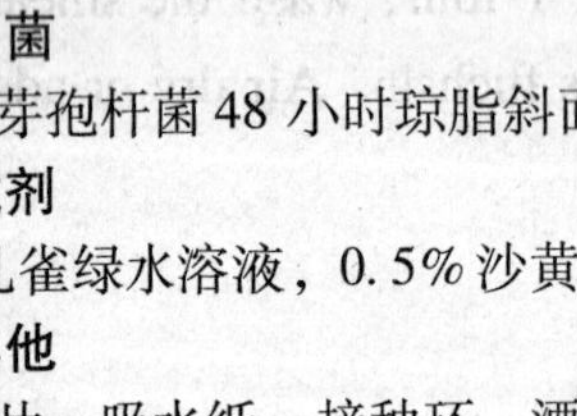

a

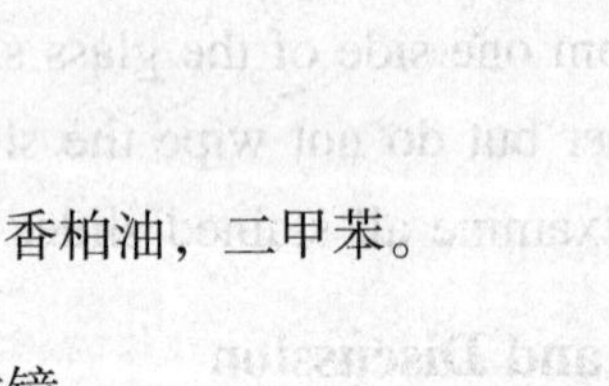

b

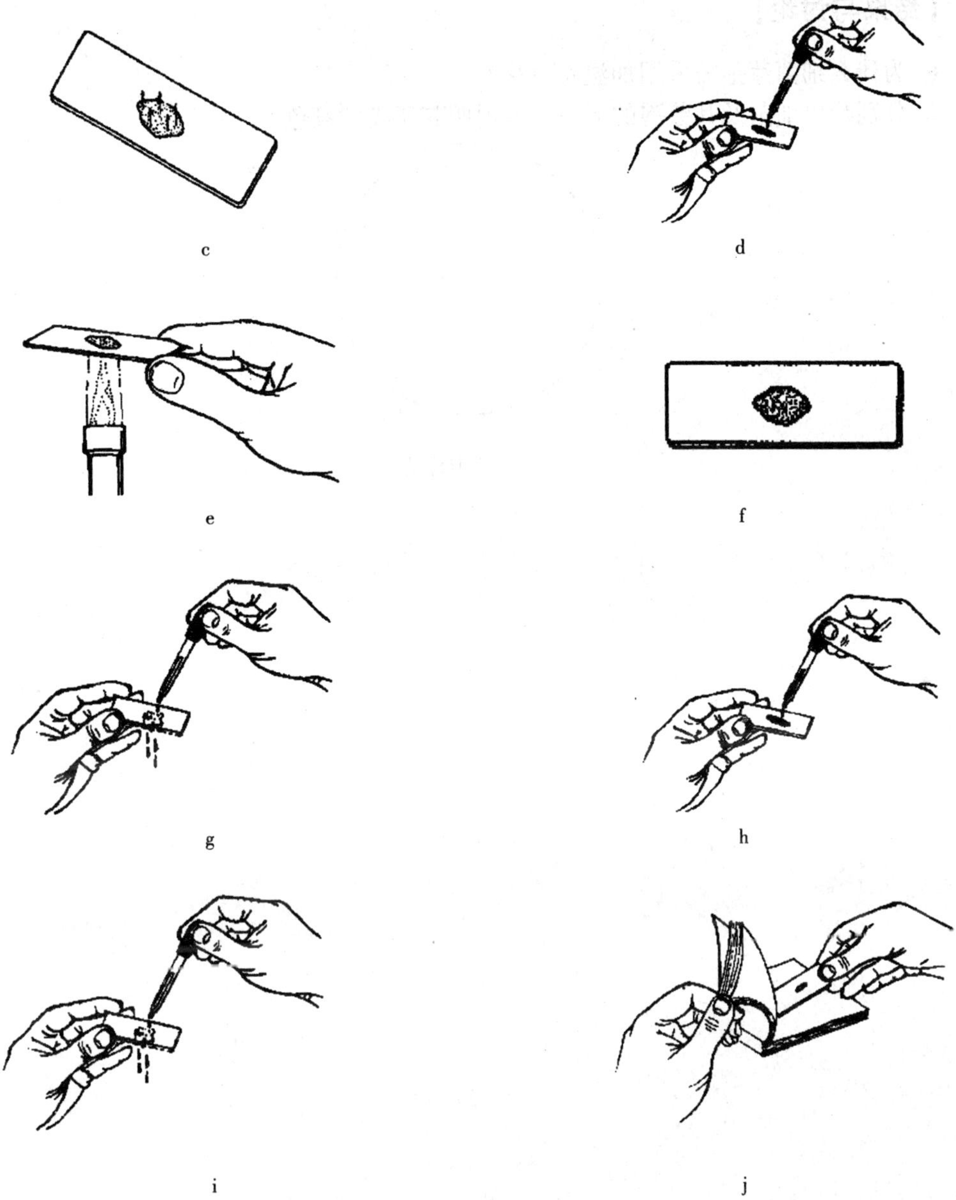

图 7－1　芽孢染色法

Fig. 7－1 Spore stain

a. 取生理盐水；　b. 涂片；　c. 干燥；　d、h. 染色；　e. 加温染色；　f. 冷却；
g、i. 冲洗；　g. 冲洗；　j. 吸干

a. Get some physiological saline;　b. Preparation of bacteria smears;　c. Air dry;　d、h. Stain;　e. Heat－stain;
f. Cool down;　g、i. Wash the smear;　j. Absorb the water

【结果与讨论】

1. 为什么细菌芽孢需采用加温染色法？
2. 分别绘出细菌及其芽孢的形态，并说明其所染的颜色。

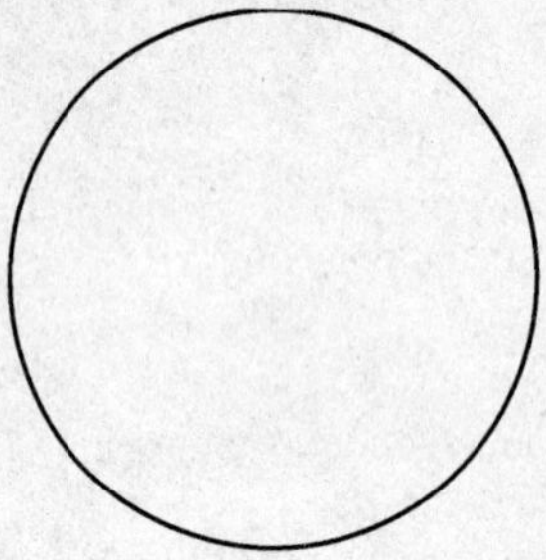

枯草芽孢杆菌

7 Spore Stain

Purposes

1. To be familiar with the procedure of the spore stain.
2. To observe the morphology of bacterial spore.

Principle

Bacterial spore structure is complex. Because of its impervious multilayers, it is difficult for bacterial spore to be stained and heat staining is required. According to the affinity of vegetative cells and spore to dyes, different colors will be obtained. Spore stain is involved in primary stain with malachite green and counterstain with safranin. At first, the bacteria are stained by malachite green. After being washed with water, only vegetative cells are decolorized because the impervious property of bacterial spore. So vegetative cells will be colorized by safranin and appear red.

Materials and Apparatus

1. Cultures

48h nutrient agar slant culture of *Bacillus subtilis*.

2. Reagents

5% malachite green, 0.5% safranin, physiological saline, cedar wood oil, xylene.

3. Apparatus

Glass slides, water – adsorption paper, inoculating loop, lamp, lens paper, microscope.

Procedures

1. Preparation of bacterial smears, air dry, and fixation refer to simple staining techniques.
2. Heat – stain it with 5% malachite green for a minute and let the smear cool.
3. Wash the smear with tap water from one side of the glass slide to remove excess stain.

4. Stain the smear with 0.5% safranin for 30s.

5. Wash the smear with tap water from one side of the glass slide to remove excess stain.

6. Air dry or adsorb water with paper, but do not wipe the slide.

7. Examine the stained slide under oil immersion.

Results and Discussion

1. Why is heating necessary in spore staining?

2. Draw a representative field for the vegetative cell and bacterial spore and indicate the color of vegetative cell and the spore.

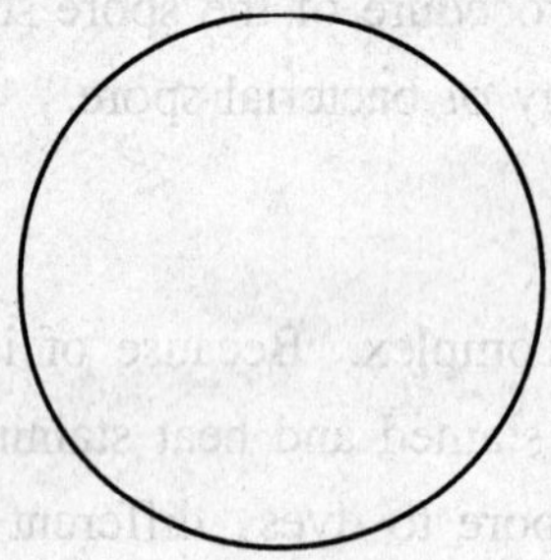

Bacillus subtilis

实验八　荚膜染色法

【实验目的】

1. 熟悉细菌荚膜的染色方法。
2. 用油镜观察细菌荚膜的形态结构。

【实验原理】

荚膜是细菌在一定营养条件下向细胞壁表面分泌的一层松散透明、黏度大的胶状物质，其组成随细菌种类而异，多数为多糖或多肽类物质。荚膜不易着色，可采用负染色法，即将菌体和背景着色，菌体周围的透明层即为荚膜。由于荚膜为非离子型，染料只附着表面而不能与其结合；同时，荚膜为水溶性物质，硫酸铜可用来洗去荚膜表面的染料而不能去除与细胞结合的染料，所以负染色后，菌体显红色，硫酸铜吸附到荚膜上而使其显亮蓝色。细菌荚膜易变形，制片时一般不用加热固定。

【实验材料】

1. 细菌

肺炎球菌培养液。

2. 动物

小白鼠。

3. 试剂

石炭酸复红染液，20% 硫酸铜溶液，生理盐水，香柏油，二甲苯。

4. 其他

注射器，载玻片，吸水纸，接种环，酒精灯，擦镜纸，显微镜。

【实验方法】

1. 取肺炎链球菌培养液注射于小白鼠腹腔中（约 0. 5ml）。
2. 小白鼠死亡后，立即取腹腔液涂片。
3. 自然干燥。
4. 石炭酸复红染液染色，并在酒精灯上加热至有蒸汽产生，开始计时，维持 3 ~ 5 分钟。
5. 以 20% 硫酸铜溶液洗涤。
6. 干燥后用油镜检查，在红色背景中可见深红色菌体，在菌体周围有无色透明圈，此为荚膜。

【结果与讨论】

1. 说明细菌荚膜染色的原理。
2. 绘出含荚膜的肺炎链球菌的形态结构。

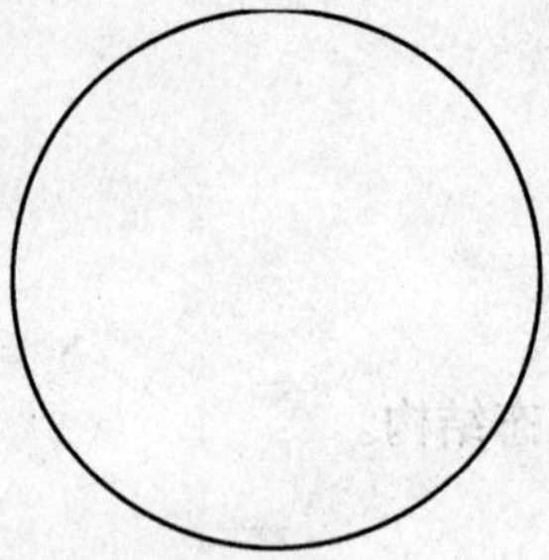

肺炎球菌

8 Capsule Stain

Purpose

1. To be familiar with the capsule stain.
2. To observe the morphology of bacterial capsule.

Principle

A capsule is a gelatinous outer layer secreted by the cell that surrounds and adheres to the cells. The composition of capsule changes with the kinds of bacteria and most of bacterial capsules are composed of polysaccharides and polypeptides. Capsule is difficult to be stained and negative stain is usually required. Because the capsule is nonionic, stain only adheres to the capsule without binding to it. In addition, the capsule is water - soluble, copper sulfate is used to wash the primary stain out of the capsular material without removing the stain that is bound to the cell wall. At the same time, it acts as a counterstain as it is absorbed into the decolorized capsular material. The capsule will now appear light blue in contrast to the deep purple color of the cell. Bacterial capsule is easy to be out of shape and heat fixation is not necessary in smear preparation.

Materials and Apparatus

1. Cultures

Culture media of *Pneumococcus pneumoniae.*

2. Animal

Mouse.

3. Reagents

Carbol fuchsin, 20% copper sulfate, physiological saline, cedar wood oil, xylene.

4. Apparatus

Glass slides, water - adsorption paper, inoculating loop, lamp, lens paper, microscope.

Procedures

1. Inject mouse (i. p.) with 0. 5ml of *Pneumococcus pneumoniae* culture.

2. After the mouse is dead, take a drop of abdomen and smear it.

3. Let the smear to air dry.

4. Stain it with carbol fuchsin and heat it until steam appears, maintain 3 ~ 5 min.

5. Wash the smear with 20% copper sulfate.

6. Air dry or adsorb water with paper but do not wipe the slide. Examine the stained slide under oil immersion.

Results and Discussion

1. What's the principle of capsule staining techniques?

2. Why must heat fixation be omitted during the preparation of smear for capsule staining?

3. Draw a representative field for *Pneumococcus pneumoniae* and indicate the color of cell, bacterial capsule and the slide background.

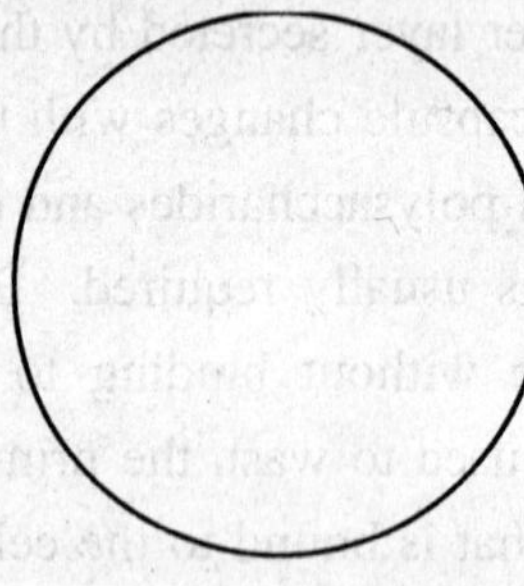

Pneumococcus pneumoniae

实验九　鞭毛染色法

【实验目的】

1. 熟悉细菌鞭毛的染色方法。
2. 用油镜观察细菌鞭毛的形态结构。

【实验原理】

鞭毛是细菌菌体表面生长的纤细、弯曲、能收缩的丝状物，其主要化学组成为蛋白质。鞭毛的长度约为 2 ~ 5μm ，最长可达 50μm ，直径很细，一般为 10 ~ 20nm 。鞭毛可采用电子显微镜来观察，也可用特殊的鞭毛染色法在油镜下观察。

【实验材料】

1. 细菌

伤寒沙门菌 8 小时菌培养液。

2. 试剂

鞭毛染色液甲（明矾饱和液 20ml，20% 鞣酸 10ml，95% 乙醇液 15ml，蒸馏水 10ml，复红酒精饱和溶液 3ml，混合而成），鞭毛染色液乙（硼砂 1g，美蓝 1g 溶于 200ml 蒸馏水中），生理盐水，香柏油，二甲苯。

3. 其他

载玻片，吸水纸，接种环，酒精灯，擦镜纸，离心机，离心试管，无菌吸管。

【实验方法】

1. 取菌液 2ml，加蒸馏水 2ml，离心沉淀（1000 转/分，10 分钟，吸出上清液，再加入蒸馏水，摇匀后离心沉淀。如此操作 3 ~ 4 次，吸出上清液，沉淀后加入蒸馏水 2 ml，摇匀后待用。
2. 用接种环在玻片上滴上述制备菌液，使流成长条。
3. 干燥（在空气中自然干燥）。
4. 加鞭毛染色液甲染色 5 分钟，水洗。
5. 加鞭毛染色液乙染色 1 ~ 2 分钟，水洗。
6. 待干燥后用油镜镜检，菌体呈蓝色，鞭毛呈红色。

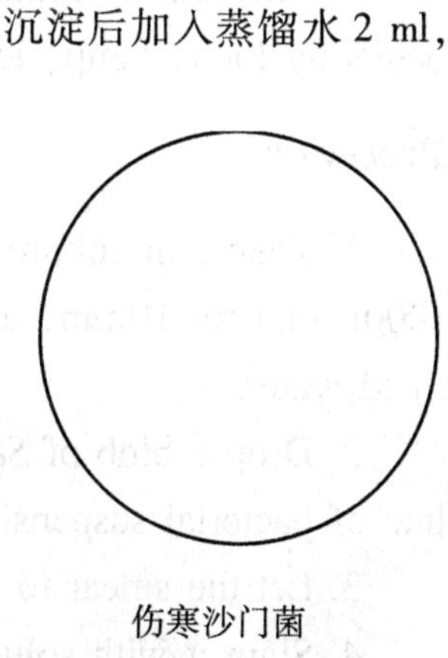

伤寒沙门菌

【结果与讨论】

1. 说明细菌鞭毛染色的原理。
2. 绘出含鞭毛的伤寒沙门菌的形态结构。

9 Flagella Stain

Purpose

1. To be familiar with the flagella stain.
2. To observe the morphology of bacterial flagella.

Principle

Flagella are fine, threadlike organelles of locomotion, composed of protein. They are slender (about 10 to 20nm in diameter and 2 to 5μm in width). Bacterial flagella can be observed with electron microscope or by light microscope after flagella staining.

Materials and Apparatus

1. Cultures

8h culture of *Salmonella typhi.*

2. Reagents

Solution A (mixtures of 20ml saturated alum solution, 10ml 20% tannic acid, 15ml 95% ethanol, 3ml fuchsin in ethanol), Solution B (1g methylene blue and 1g sodium borate dissolved in 200ml distilled water), physiological saline, cedar wood oil, xylene.

3. Apparatus

Centrifugal machine, centrifugal tube, glass slides, water - adsorption paper, inoculating loop, lamp, lens paper, microscope.

Procedures

1. Take 2ml culture media of *Salmonella typhi* in 2ml distilled water, centrifuge at 1000r/min for 10min, and discard the upper phase. Suspend the precipitate in 2ml distilled water.

2. Drop a blob of *Salmonella typhi* suspension solution on a glass slide and form a line of bacterial suspension solution.

3. Let the smear to air dry.

4. Stain it with solution A for 5min, wash the smear with tap water from one side

of the glass slide to remove excess stain.

5. Stain it with solution B for 1 to 2 min, wash the smear with tap water from one side of the glass slide to remove excess stain.

6. Air dry or adsorb water with paper but do not wipe the slide. Examine the stained slide under oil immersion lens.

Results and Discussion

1. What's the principle of flagella staining techniques?

2. Draw a representative field for *Salmonella typhi.*

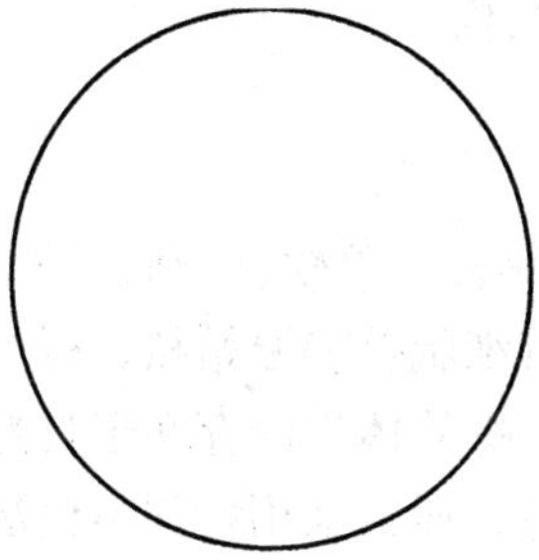

Salmonella typhi

实验十　螺旋体的染色和形态观察

【实验目的】

1. 掌握螺旋体的染色方法。
2. 用油镜观察螺旋体的形态结构。

【实验原理】

螺旋体（spirochete）是一类细长、柔软、弯曲呈螺旋状、运动活泼的原核细胞型微生物。其基本结构与细菌类似，有细胞壁和原始核，行二分裂法繁殖，对抗生素敏感。螺旋体无鞭毛，其运动有赖于轴丝。螺旋体广泛存在于自然界和动物体内，种类繁多。

螺旋体对革兰染色法不易着色，通常采用特殊染色法（如镀银法），油镜观察。

【实验材料】

1. 染料与其他试剂

方他那染液（甲、乙、丙液），生理盐水，香柏油，二甲苯。

2. 其他

干净载玻片，牙签，酒精灯，擦镜纸，显微镜。

【实验方法】

1. 以牙签挖取齿垢置于玻片上，涂成薄膜，标本自行干燥。
2. 加方他那甲液固定 1 分钟，水洗。
3. 加方他那乙液，加热至蒸汽发生，染色 30 秒，稍冷，水冲洗。
4. 加方他那丙液后加热至蒸汽发生，染色 30 秒，稍冷。
5. 自然干燥或吸干，油镜检查，染色后，背景为黄色，螺旋体及细菌呈棕色。

【结果与讨论】

1. 说明螺旋体染色的原理。
2. 绘制疏螺旋体的形态结构。

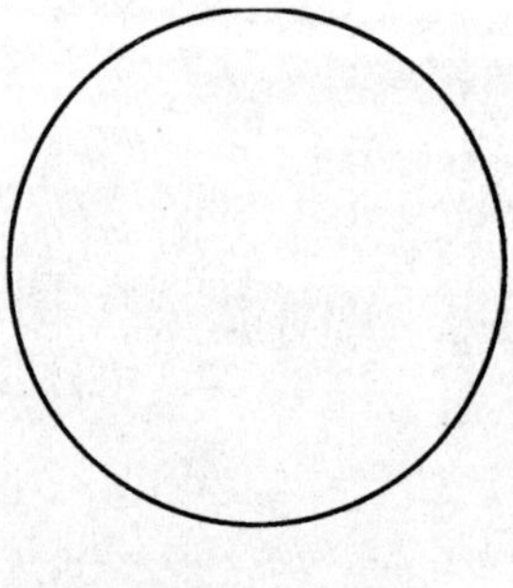

疏螺旋体的形态结构

10 Spirochetes Morphology

Purpose

1. To be familiar with the spirochetes stain.
2. To observe the morphology of spirochetes.

Principle

Spirochetes are slender, soft, curved, movable, and helical prokaryotic cells which are about 5 to 10nm in length and 0. 2μm in width. In addition to dupilication, the primary structure of spirochetes is similar to bacteria with cell wall and nucleoid. Spirochetes are sensitive to antibiotics and can move caused by a special axial filament other than flagellar rotation.

Materials and Apparatus

1. Reagents

Fontana Solution A, solution B, solution C, physiological saline, cedar wood oil, xylene.

2. Apparatus

Glass slides, water – adsorption paper, toothpick, lamp, lens paper, microscope.

Procedures

1. Drop a blob of physiological water on the center of glass slide, take some tartar with toothpick into the physiological water on the glass slide and smear it, let the smear to air dry.

2. Fix it with solution A for one minute, wash the smear with tap water from one side of the glass slide to remove excess stain.

3. Heat – stain it with solution B for 30s, cool and wash the smear with tap water from one side of the glass slide to remove excess stain.

4. Heat – stain it with solution C for 30s, cool and wash the smear with tap water from one side of the glass slide to remove excess stain.

5. Air dry or adsorb water with paper but do not wipe the slide. Examine the

stained slide under oil immersion lens.

Results and Discussion

1. What's the principle of spirochetes staining techniques?
2. Draw a representative field for spirochetes.

Spirochetes

实验十一　悬滴法观察细菌的动力

【实验目的】

1. 了解不染色标本检查法。
2. 用悬滴法观察活细菌及其动力。

【实验原理】

细菌的鞭毛是细菌的运动器官，细菌是否具有鞭毛是细菌分类鉴定的依据之一。有鞭毛的细菌在液体中能发生位置上的变化而作真正的运动；而无鞭毛的细菌由于水分子的撞击，在原位颤动，称之为布朗运动。可采用悬滴法直接在普通光学显微镜下观察细菌是否具有运动能力，以此来判定细菌是否具有鞭毛。

悬滴法是将细菌培养液滴加在洁净的盖玻片中央，在其周围涂上凡士林，将盖玻片置于凹玻片中央，使盖玻片上的液滴悬于凹槽上方（图 11 －1），在光学显微镜下观察细菌的运动情况。

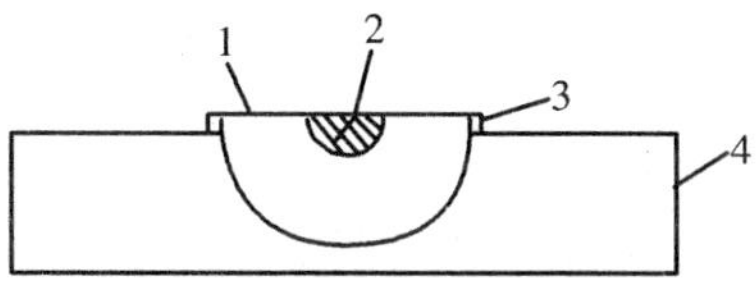

图 11 －1　悬滴法示意图

Fig. 11 －1 Hanging drop method

1. 盖玻片；　2. 菌液；　3. 凡士林；　4. 凹玻片

1. Cover slip；　2. Bacteria liquid culture；　3. Vaselin；　4. Concavity Slide

【实验材料】

1. 细菌

枯草芽孢杆菌和金黄色葡萄球菌的 8 小时液体培养物。

2. 其他

凹玻片，盖玻片，牙签，接种环，酒精灯，凡士林，显微镜。

【实验方法】

1. 悬滴标本的制备

（1）用灭菌的接种环挑取细菌液体培养物一环，加于平放桌上之洁净盖玻片中央。

（2）在洁净凹玻片的凹面周围用牙签涂上少许凡士林（或浆糊）。

（3）将凹玻片凹面向下，覆盖在已滴加有菌液的盖玻片上，轻压，使盖玻片粘在凹玻片上。

（4）迅速反转凹玻片，使盖玻片上之菌液悬于凹玻片之凹面中。

2. 显微镜检查

（1）将制备好的悬滴标本固定于载物台上。

（2）先用低倍镜对光（光圈缩小，聚光器下降，使光线较暗），找到菌液之边缘，并调至视野中央。

（3）转换高倍镜，再轻轻调节细调节器，即可看到透明的菌体。观察两种细菌的不同运动特征。

【结果与讨论】

1. 分别描述金黄色葡萄球菌和枯草芽孢杆菌的运动特征。
2. 检测细菌是否具有动力的方法有哪些？

11 Observation of Bacterial Motion with Hanging Drop

Purpose

1. To be familiar with the preparation of unstained specimen.
2. To observe the motion of living bacteria.

Principle

Flagella are motion organelles of bacteria. Bacteria with flagella can move from one place to another in culture solution. On the other hand, bacteria without flagella will vibrate due to water collision. That is called Brown's motion. Here hanging drop method is used to observe the motion of living bacteria with bright – field microscope. As shown in Fig. 11 – 1, a drop of bacteria liquid culture is suspended on the center of cover slip.

Materials and Apparatus

1. Cultures

8h culture media of *Staphylococcus aureus* and *Bacillus subtilis.*

2. Apparatus

Concavity slide, cover slip, toothpick, inoculating loop, lamp, lens paper, microscope.

Procedures

1. Preparation of specimen

(1) Place a drop of culture media of *Staphylococcus aureus* and *Bacillus subtilis* respectively on a glass slide and mix it with inoculating loop.

(2) Smear some vaseline around the edges of the cover slip with toothpick.

(3) Place a concavity slide on the cover slide and press it softly for sticking.

(4) Reverse the specimen rapidly.

2. Observation with microscope

(1) Place the specimen on the stage.

(2) Observe it with low power lens (10 ×) and fix the edge of drop.

(3) Observe it with high power lens (40×).

Results and Discussion

1. Describe the motive characteristics of *Staphylococcus aureus* and *Bacillus subtilis.*

2. How to decide whether bacteria can move?

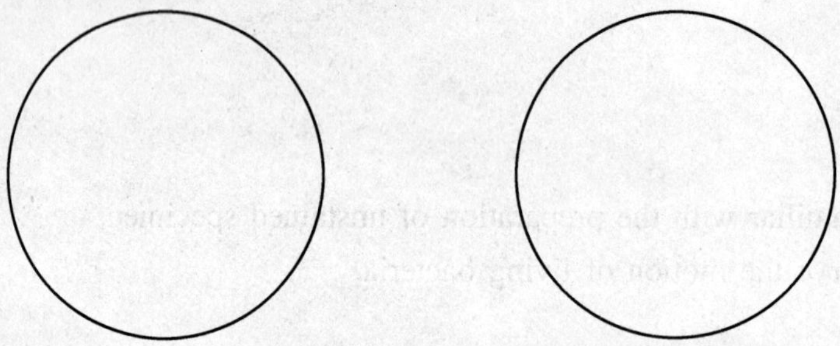

Staphylococcus aureus *Bacillus subtilis*

实验十二　酵母菌的形态结构观察

【实验目的】

1. 熟悉酵母菌的单染色法。
2. 观察酵母菌的形态结构。
3. 熟悉酵母菌子囊孢子的观察方法。

【实验原理】

酵母菌是单细胞真菌，有圆形、卵圆形等形态，无性繁殖主要以芽殖或裂殖的方式进行，有性繁殖时形成子囊孢子。美蓝是一种无毒性染料，其氧化型呈蓝色，还原型为无色。由于酵母菌的活细胞新陈代谢的作用，细胞内具有较强的还原能力，经染色的酵母菌活细胞能使美蓝从蓝色的氧化型变为无色的还原型而呈无色；由于死细胞或代谢缓慢的细胞则因无还原能力或还原能力弱，从而被染成蓝色或淡蓝色。

酵母菌在其适宜的生长条件下可形成子囊孢子，葡萄糖－醋酸盐培养基特别适用于酿酒酵母子囊孢子的形成。

【实验材料】

1. 菌种

酿酒酵母改良沙氏琼脂培养基 48 小时斜面培养物，葡萄糖－醋酸盐培养基的 24 小时斜面培养物。

2. 试剂

0.05% 和 0.1% 碱性美蓝染液，石炭酸复红染液，3% 酸性酒精，孔雀绿芽孢染液，生理盐水，香柏油和二甲苯。

3. 其他

显微镜，载玻片，盖玻片，擦镜纸，酒精灯，显微镜。

【实验方法】

1. 酵母菌细胞美蓝染色法

（1）在载玻片中央滴加一滴 0.1% 碱性美蓝染液，按无菌操作法用接种环在改良沙氏琼脂培养基斜面上培养 48 小时的酿酒酵母少许，于美蓝染液中混匀。

（2）用镊子夹一块盖玻片，先将盖玻片的一边与液滴接触，然后将整个盖玻片慢慢放下，同时避免气泡产生。标本片静置 3 分钟。

（3）先用低倍镜观察，再用高倍镜观察酿酒酵母的形态和芽殖情况，根据细胞是否染色来区别活细胞和死细胞。

2. 酵母菌子囊孢子染色法

（1）用接种环挑取葡萄糖－醋酸盐培斜面养基上的酿酒酵母于载玻片上，涂片、干燥和固定。

（2）加石炭酸复红染液，在火焰上加温染色5~10分钟（避免沸腾）。

（3）3%酸性酒精洗30~60秒，水洗。

（4）吕氏碱性美蓝数滴复染数秒钟后，水洗。

（5）油镜观察，酵母菌子囊孢子为红色，菌体为青色。如用孔雀绿芽孢染液单染（不加热），油镜观察可见绿色的子囊孢子。

【结果与讨论】

1. 为什么可用染色法区别酵母菌活细胞和死细胞？
2. 绘制酵母菌菌体和子囊孢子的形态结构。

酵母菌菌体

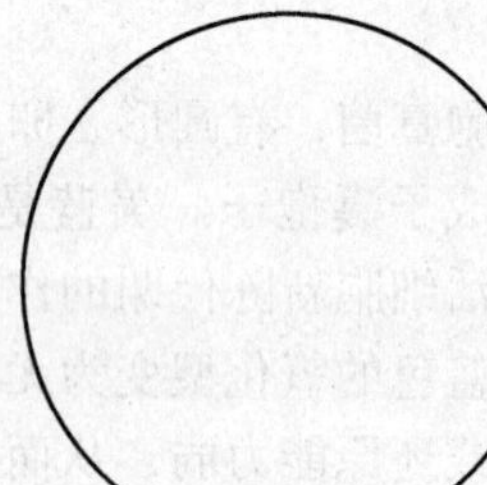

酵母菌子囊孢子

12 Yeast Morphology

Purpose

1. To become familiar with the single staining of yeast.
2. To observe the morphology of yeast cells.
3. To become familiar with ascospores staining technique of yeast.

Principle

Yeasts are unicellular organisms. Morphologically, they may be ellipsoidal, spherical, and cylindrical in some cases. Most of yeasts reproduce asexually by a process called budding, and some yeasts may also undergo sexual reproduction by forming ascospores.

Methylene blue is a nonpoison dye which appears blue in oxidation type and no color in reduction type. Viable yeast cells have the possibility of reduction due to metabolism. After stained by methylene blue, viable yeast cells reduce it to reduction type other than dead cells and appear uncolored.

Yeast cells can form ascospores under appropriate conditions and glucose - acetate media is often used as the culture of Saccharomyces for forming ascospores.

Materials and Apparatus

1. Cultures

48h Sabouraud agar culture of *Saccharomyces cerevisiae.* 24h glucose - acetate agar culture of *Saccharomyces cerevisiae.*

2. Reagents

0.05% and 0.1% methylene blue, safranin, 3% acidic ethanol, 0.5% malachite green, physiological saline, cedar wood oil, xylene.

3. Apparatus

Glass slides, tissue, inoculating loop, lamp, lens paper, microscope.

Procedures

1. Yeast stain with methylene blue

(1) Drop a blob of methylene blue on the center of glass slide.

(2) Take some yeast from sabouraud agar culture with inoculating loop and suspend a loopful of yeasts culture into a few drops of methylene blue solution.

(3) Cover the suspension slowly with a cover slip and stay it for 3 min.

(4) Examine the slide with microscope under low and high power.

2. Ascospores stain

(1) Smear 24h glucose - acetate agar culture of *Saccharomyces cerevisiae* on glass slide, air dry and fixation refer to simple staining techniques.

(2) Heat stain the smear with safranin for 5 to 10min..

(3) Wash the smear with 3% acidic ethanol for 30 ~ 60 s, then wash with tap water to remove excess ethanol.

(4) Stain the smear with methylene blue for some seconds and wash with tap water.

(5) Examine the stained slide under oil immersion.

Results and Discussion

1. Why can methylene blue stain be used to distinguish viable and dead cell of yeast?

2. Draw a representative field for yeast under low - power and high - power magnifications.

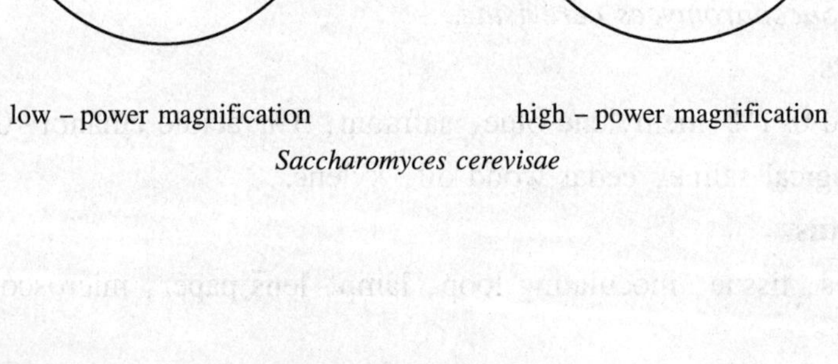

low - power magnification high - power magnification

Saccharomyces cerevisae

实验十三　微生物细胞大小测定

【实验目的】

1. 了解目镜测微尺和镜台测微尺的构造和使用原理
2. 掌握微生物细胞大小的测定方法。

【实验原理】

微生物细胞的大小是微生物重要的形态特征之一，由于菌体很小，只能在显微镜下测量。用于测量微生物细胞大小的工具有目镜测微尺和镜台测微尺。

目镜测微尺是一块圆形玻片，在玻片中央把5mm长度刻成50等分，或把10 mm长度刻成100等分。测量时，将其放在接目镜中的隔板上（此处正好与物镜放大的中间像重叠）来测量经显微镜放大后的细胞物像。由于不同目镜、物镜组合的放大倍数不相同，目镜测微尺每格实际表示的长度也不一样，因此目镜测微尺测量微生物大小时须先用置于镜台上的镜台测微尺校正，以求出一定放大倍数下，目镜测微尺每小格所代表的相对长度。

镜台测微尺是中央部分刻有精确等分线的载玻片，一般将1mm等分为100格，每格长10μm（即0.01mm），专门用来校正目镜测微尺。校正时，将镜台测微尺放在载物台上，由于镜台测微尺与细胞标本处于同一位置，要经过物镜和目镜的两次放大成像才能进入视野，即镜台测微尺随着显微镜总放大倍数的放大而放大，因此从镜台测微尺上得到的读数就是细胞的真实大小，用镜台测微尺的已知长度在一定放大倍数下校正目镜测微尺，即可求出目镜测微尺每格所代表的长度，然后移去镜台测微尺，换上待测标本片，用校正好的目镜测微尺在同样放大倍数下测量微生物大小（图13－1）。

【实验材料】

1. 细菌

酿酒酵母（*Saccharomyces cerevisiae*）斜面菌种、枯草芽孢杆菌（*Bacillus subtilis*）染色标本片。

2. 器材

显微镜、目镜测微尺、镜台测微尺、盖玻片、载玻片、滴管、二重瓶、擦镜纸。

【实验方法】

1. 目镜测微尺的校正

把目镜的上透镜旋下，将目镜测微尺的刻度朝下轻轻地装入目镜的隔板上，把镜台测微尺置于载物台上，刻度朝上。先用低倍镜观察，对准焦距，视野中看清镜台测微尺的刻度后，转动目镜，使目镜测微尺与镜台测微尺的刻度平行，移动推动器，使两尺重叠，再使两尺的“0”刻度完全重合，定位后，仔细寻找两尺第二个完全重合的刻度，计数两重合刻度之间目镜测微尺的格数和镜台测微尺的格数。镜台测微尺的刻度每格长 l0μm，由下列公式算出目镜测微尺每格所代表的长度。

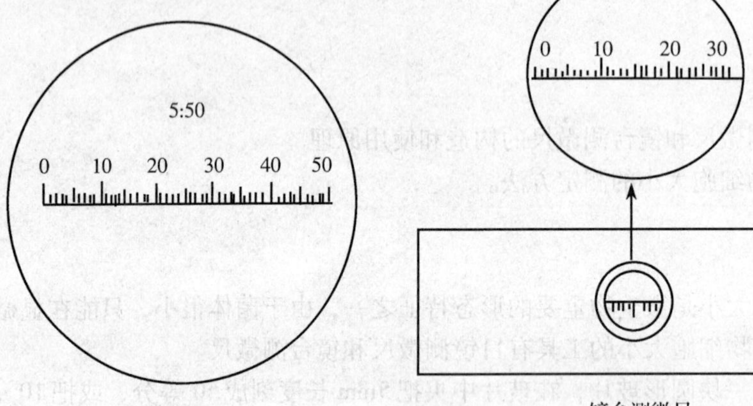

目镜测微尺
the ocular micrometer

镜台测微尺
the stage micrometer

图 13－1 目镜测微尺和镜台测微尺

Fig. 13－1 The ocular micrometer and stage micrometer

目镜测微尺每格长度（μm）＝10 n/m

式中 n 为两重合线间镜台测微尺格数；

m 为两重合线间接目测微尺格数。

例如目镜测微尺 5 小格正好与镜台测微尺 5 小格重叠，已知镜台测微尺每小格为 l0μm，则目镜测微尺上每小格长度为＝5×10 μm/5＝10μm。用同法分别校正高倍镜下和油镜下目镜测微尺每小格所代表的长度。

由于不同显微镜及附件的放大倍数不同，因此校正目镜测微尺必须针对特定的显微镜和附件（特定的物镜、目镜、镜筒长度）进行，而且只能在特定的情况下重复使用，当更换不同放大倍数的目镜或物镜时，必须重新校正目镜测微尺每一格所代表的长度。

2. 细胞大小的测定

（1）将酵母菌斜面制成一定浓度的菌悬液。

（2）取 1 滴酵母菌菌悬液制成水浸片。

（3）移去镜台测微尺，换上酵母菌水浸片，先在低倍镜下找到目的物，然后在高倍镜下用目镜测微尺测量酵母菌菌体的长，宽各占几格（不足一格的部分估计到小数点后一位数）。测出的格数乘上目镜测微尺每格的校正值，即等于该菌的长和宽。一般测量菌体的大小要在同一个标本片上测定 10～20 个菌体，求出平均值，才能代表该菌的大小。而且一般是用对数生长期的菌体进行测定。

（4）同法用油镜测定枯草芽孢杆菌染色标本的长和宽。

【结果与讨论】

1. 将实验结果填入下列表格。

表 13－1　目镜测微目尺校正结果

物镜	目尺格数	台尺格数	目尺校正值（μm）
10×			
40×			
100×			

表 13－2　酵母菌大小测定记录（格）

	1	2	3	4	5	6	7	8	9	10	11	12	13	14	15	平均值
细胞数																
长																
宽																

表 13－3　枯草芽孢杆菌大小测定记录（格）

	1	2	3	4	5	6	7	8	9	10	11	12	13	14	15	平均值
细胞数																
长																
宽																

2. 结果计算

（1）长（μm）＝平均格数×校正值

（2）宽（μm）＝平均格数×校正值

（3）大小表示：宽（μm）×长（μm）

13 Determination of the Size of Microorganisms

Purpose

1. To be familiar with the structure and principle of the ocular micrometer and stage micrometer.

2. To know how to determine the size of microorganism.

Principle

The size of microorganism is one of the important morphological characteristics. As the cell is very small, it can only be measured in the microscope. Usually, we use ocular micrometer and stage micrometer.

Ocular micrometer is a circular slide, in which center there is a length of 5mm carved into 50 equation. When measured, we placed it on the eyepiece of the partition to determination the cell images amplified by the microscope. As eyepiece, magnification lens combinations are different, each equation of the eyepiece micrometer ruler - represent different actual length. As a result, before we determine the size of the microorganism, we must correct the ocular micrometer to determine the relative length each equation represents under a certain magnification.

Stage micrometer is a glass slide in which center there are accurate bisectors, usually dividing 1mm into 100 parts. Each part is 10μm, which is special to correct ocular micrometer. When corrected, we put the stage micrometer on the object stage.

As the stage micrometer and cell samples are in the same location, their images are magnified twice going through the objective lens and eyepiece to enter the field of vision. The stage micrometer is magnified with the total magnification of the microscope. Thus, the readings of the stage micrometer are the real size of the cells. As a result, we use the known length of the stage micrometer to correct the ocular micrometer under certain magnification times to determine the length each part represent. Then remove the stage micrometer, put on the sample, and determine the size of microorganism under the same magnification times.

Materials and materials

1. Cultures

24h YPT agar slant cultures of *Saccharomyces cerevisiae* and stained specimen of *Bacillus subtilis.*

2. Apparatus

Microscope, ocular micrometer, stage micrometer, microscopic glass, glass slide, dropping tube, double bottle, lens paper.

Procedures

1. Correct the ocular micrometer

(1) Load ocular micrometer and put the stage micrometer on the object stage.

(2) Observe with low - power objective lens.

(3) When the ocular micrometer's graduation is clear, roll the eye lens, force the ocular micrometer paralleled with the stage micrometer.

(4) Overlap the "0" of two micrometer, and then find the second scale overlapping. Count the number of the lattices between the two overlapping scale.

(5) Calculate

The length of each lattice in ocular micrometer (μm) = $10n/m$

n: the number of lattices between the two overlapping scale in stage micrometer

m: the number of lattices between the two overlapping scale in ocular micrometer

When observe with high - power or oil objective lens, the correct methods are the same.

2. Determination of the size of the cells

(1) Prepare the cells suspension of the yeast.

(2) Prepare the slide with a drop of cells suspension.

(3) Remove the ocular micrometer, replace it with the slide. find the object with low - power objective lens, then determine the size (length and width) and under the high - power objective lens.

(4) The size of the microorganism = The number of the lattices of length/ width × the correction value of the ocular micrometer

(5) Determine 10 - 20 cells and calculate the average.

(6) The microorganism often be prepared in exponential phase.

(7) Determine the length and the width of *Bacillus subtilis* under the oil objective lens with the same method.

Results and Discussion

(1) Filling the experimental results into the following tables.

Table 13 – 1 Correction of ocular micrometer

Objective lens	Lattices of ocular micrometer	Lattices of stage micrometer	Correction value (μm)
10 ×			
40 ×			
100 ×			

Table 13 – 2 Size of yeast (number of lattices)

	1	2	3	4	5	6	7	8	9	10	11	12	13	14	15	Average
Cells number																
Length																
Width																

Table 13 – 3 Size of *Bacillus subtilis* (number of lattices)

	1	2	3	4	5	6	7	8	9	10	11	12	13	14	15	Average
Cells number																
Length																
Width																

(2) Calculation of the results

length (μm) = average × correction value

Width (μm) = average × correction value

Size: length (μm) × width (μm)

实验十四　放线菌、霉菌的菌丝和孢子形态观察

【实验目的】

1. 熟悉压片法和小培养法观察菌丝和孢子的形态。
2. 观察放线菌、霉菌的菌丝和孢子形态。

【实验原理】

放线菌是一类原核单细胞型微生物，细胞的基本结构类似于细菌。放线菌具有丝状分枝，形成基内菌丝、气生菌丝和顶端部分分化而成的孢子丝，孢子丝断裂形成孢子。孢子丝可呈螺旋状、波浪状或分枝状等，有的还能分泌水溶性色素到培养基内。孢子呈圆形、杆型和椭圆型等，且具有各种颜色。气生菌丝及孢子的形态和颜色常作为放线菌分类和鉴定的重要依据。

霉菌为真核细胞型微生物，大多数为多细胞，具有分枝状菌丝体，菌丝较粗，孢子有各种不同的形状，可分为有性孢子和无性孢子。

放线菌、霉菌菌丝和孢子形态可采用印片法和小培养法来观察，其中小培养法可分为凹玻片法和插片法。

【实验材料】

1. 细菌

放线菌和霉菌培养物。

2. 染料

美蓝染色液。

3. 器材

盖玻片、载玻片、凹玻片、显微镜。

【实验方法】

1. 印片法观察放线菌和霉菌的菌丝和孢子形态

（1）用盖玻片在放线菌（抗生素产生菌 2809）和霉菌（青霉）菌落表面轻轻一按，即印取孢子。

（2）将印有孢子的一面朝下，放在滴有一滴美蓝染液的载玻片上，静置 3 分钟，使放线菌和霉菌孢子可着色。

（3）用油镜观察放线菌和霉菌孢子、孢子丝的形态结构。

2. 凹玻片法观察放线菌和霉菌的菌丝和孢子形态

（1）取一灭菌的凹玻片，于凹窝内滴加少量熔化的固体培养基，待凝。

（2）接种少许菌种，覆盖一灭菌的盖玻片，盖玻片与凹窝间稍留一点空隙以便通气。

（3）将凹玻片置无菌平皿内，皿内垫一块潮湿的脱脂棉以维持湿度，25～28℃恒温培养3天。观察生长状况。

3. 盖玻片法观察放线菌和霉菌的菌丝和孢子形态

（1）将菌种均匀的涂布在固体平板培养基上。

（2）用镊子夹一张无菌盖玻片斜插入平板内的培养基中，插入深度为盖玻片高度的二分之一或三分之一。

（3）25～28℃恒温箱中培养5～7天。

（4）用镊子取出盖玻片放在载玻片上，用低倍镜或高倍镜观察菌丝及孢子的情况。

【结果与讨论】

1. 分别绘出放线菌、霉菌的孢子、孢子丝和菌丝的形态特征。

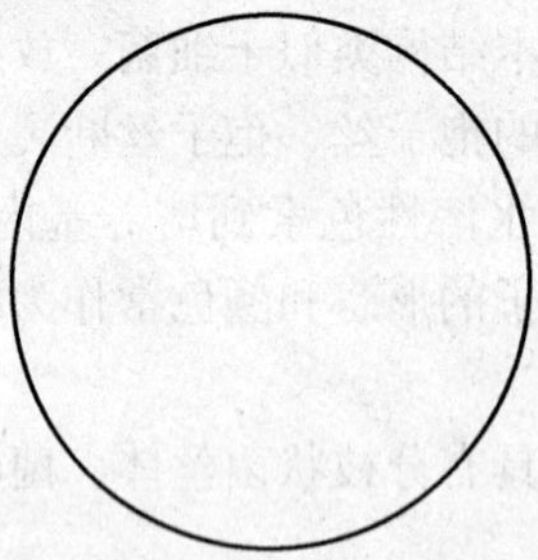

放线菌

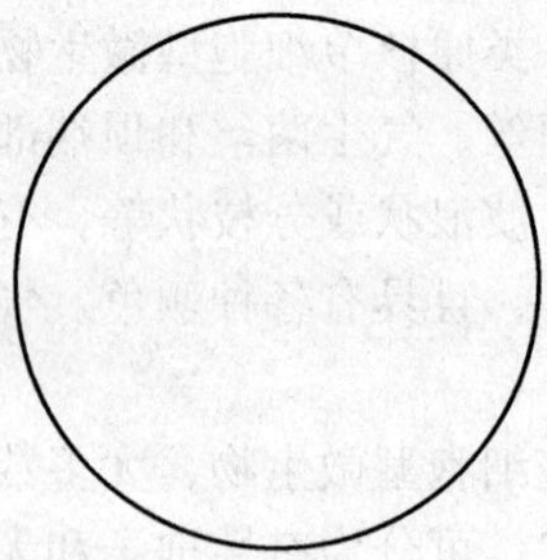

霉菌

2. 放线菌、霉菌的菌丝和孢子有何区别?

14 Morphology of Mycelia and Spore for Actinomycetes and Molds

Purpose

1. To be familiar with the observation method of mycelia and spore.
2. To observe the morphology of mycelia and spore for actinomycetes and molds.

Principle

Actinomycetes are prokaryotic and single cellular microorganisms. The primary cell structure of actinomycetes is similar to bacteria. Actinomycetes have filament which contains substrate filament, aerial filament and sporebearing filament. Sporebearing filament differentiate to spore. The shapes of sporebearing filament are sphere, ellipse and rod. The morphology and color of aerial filament and spore are important for the classification and characterization of actinomycetes.

Molds are eukaryotic cells and most of them are multicellular and filamented. The filament of molds is often wider than that of actinomycetes. Spores have different shapes and are divided into asexual or sexual ones.

The morphology of mycelia and spore for actinomycetes and molds can be observed by microscope and cultivated on concavity slide or cover slip.

Materials and Apparatus

1. Microorganisms

Cultures of actinomycetes and molds.

2. Reagent

Methylene blue.

3. Apparatus

Glass slides, inoculating loop, lamp, lens paper, microscope.

Procedures

1. Observation of mycelia and spore of actinomycetes and molds by microscope

(1) Press a cover slip on the colony surface of actinomycetes softly to stick some spore.

(2) Place a drop of methylene blue on a glass slide. Reverse the cover slip and place it on the glass slide and stay it for 3 min to let spore be stained.

(3) Observe the morphology of mycelia and spore under oil immersion.

2. Observation of mycelia and spore of actinomycetes and molds by cultivation on concavity slide

(1) Place some agar media into the concavity of sterilized concavity slide and cool.

(2) Inoculate actinomycetes or molds with inoculating loop and place a cover slip on the concavity. Note that a little space should be kept for air.

(3) Place the concavity slide and wet cotton into a sterilized plate and culture at 25 ~ 28℃ for 3 days. Record the culture state.

3. Observation of mycelia and spore of actinomycetes and molds by cultivation on cover slide

(1) Inoculate actinomycetes and molds on the culture plate.

(2) Place a sterilized cover slip into culture media at 45℃ and 1/2 or 1/3 height.

(3) Culture at 25 ~ 28℃ for 5 ~ 7 days.

(4) Grip the cover slip with tweezers and place it on the glass slide. Observe the morphology of mycelia and spore with microscope (low power or high power).

Results and Discussion

1. Describe the morphological characteristics of the mycelia and spore of actinomycetes and molds.

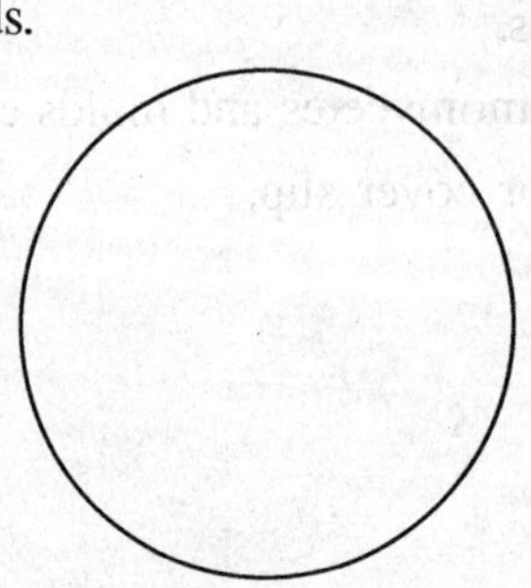

actinomycetes

molds

2. What's the difference of the mycelia and spore between actinomycetes and molds?

实验十五　细菌、放线菌、酵母菌和霉菌的菌落特征观察

【实验目的】

通过观察培养平板，了解细菌、放线菌、酵母菌和霉菌的菌落特征。

【实验原理】

菌落是将微生物接种于固体平板培养基上，经恒温培养后由单个细胞形成的集落。菌落特征决定于组成菌落的细胞结构和生长行为，不同的微生物具有不同的菌落特征，用以对微生物进行初步鉴定。

【实验材料】

1. 菌种

金黄色葡萄球菌，藤黄八叠球菌、大肠埃希菌、枯草芽孢杆菌、放线菌、酿酒酵母和霉菌（青霉、曲霉、毛霉、根霉）的斜面和平板培养物。

2. 培养基

普通琼脂平板，淀粉琼脂平板，改良沙氏琼脂平板。

【实验方法】

1. 将金黄色葡萄球菌，藤黄八叠球菌、大肠埃希菌、枯草芽孢杆菌接种于普通琼脂平板培养基，37℃恒温培养 24 小时。
2. 将放线菌接种于淀粉琼脂平板培养基，28℃恒温培养 5 ~7 天。
3. 将酿酒酵母菌接种于改良沙氏琼脂平板培养基，30℃恒温培养 48 小时。
4. 将青霉、曲霉、毛霉、根霉接种于改良沙氏琼脂平板培养基，25℃恒温培养 1 ~2 周。
5. 菌落特征观察。

【结果与讨论】

1. 细菌的菌落特征

特征＼细菌	金黄色葡萄球菌	藤黄八叠球菌	大肠埃希菌	枯草芽孢杆菌
大小				
形态				
表面				
边缘				
颜色				
可溶性色素				

2. 放线菌的菌落特征

特征＼放线菌	放线菌1	放线菌2	放线菌3	放线菌4	放线菌5
大小					
形态					
表面（崎岖、皱褶或平滑）					
气生菌丝颜色及状态（粉状、绒粉状或短毛状）					
孢子丝与孢子堆的颜色					
基内菌丝的颜色					
可溶性色素					

3. 真菌的菌落特征

特征＼真菌	青霉	曲霉	毛霉	根霉	酵母菌
大小					
形态					
表面（崎岖、皱褶或平滑）					
气生菌丝颜色及状态（粉状、绒粉状或短毛状）					
孢子丝与孢子堆的颜色					
基内菌丝的颜色					
可溶性色素					

4. 试比较细菌、放线菌、酵母菌和霉菌的菌落特征有何异同。

15 Colony Characteristics of Bacteria, Actinomycetes, Yeast and Molds

Purpose

1. To observe the colony morphology of bacteria, actinomycetes, yeast and molds.
2. To distinguish the colony of bacteria, actinomycetes, yeast and molds.

Principle

After inoculated on agar culture media and cultured at approximate temperature for some time, microorganisms grow to form colonies from single cell which can been seen by the naked eye. The characteristics of colonies depends on their cell structure and growth behavior which is further used to the identification of microorganisms.

Materials and Apparatus

1. Microorganisms

Staphylococcus aureus, *Sarcina lutea*, *Escherichia coli*, *Bacillus sublitis*, *actinomycetes*, *Saccharomyces cerevisiae*, molds (Penicillium, Aspergillus, Mucor, Rhizopus).

2. Media

Nutrient agar slant, starch agar slant, Sabouraud agar slant.

Procedures

1. Inoculating *Staphylococcus aureus*, *Sarcina lutea*, *Escherichia coli*, *Bacillus sublitis* on nutrient agar slant and culture at 37℃ for 24h.
2. Inoculating actinomycetes on starch agar slant and culture at 28℃ for 5 to 7 days.
3. Inoculating *Saccharomyces cerevisiae* on sabouraud agar slant and culture at 30℃ for 48h.
4. Inoculating molds on Sabouraud agar slant and culture at 25℃ for 1 to 2 weeks.
5. To observe the colony morphology of bacteria, actinomycetes, yeast and molds.

Results and Discussion

1. Describe the characteristics of bacterial colonies.

Character \ Bacteria	*S. aureus*	*S. lutea*	*E. coli*	*B. sublitis*
Size				
Shape				
Surface				
Edge				
Color				
Dissolved pigment				

2. Describe the characteristics of actinomycetes colonies.

Character \ Actinomycetes	No. 1	No. 2	No. 3	No. 4	No. 5
Size					
Shape					
Surface					
Shape and color of aerial mycelia					
Color of spore mycelia					
Color of substrate mycelia					
Dissolved pigment					

3. Describe the characteristics of fungi.

Character \ Fungi	Penicillum	Aspergillus	Mucor	Rhizopus	Sacchariomyces
Size					
Shape					
Surface					
Shape and color of aerial mycelia					
Color of spore mycelia					
Color of substrate mycelia					
Dissolved pigment					

4. Compare the colony characteristics of bacteria, actinomycetes, yeast and molds.

第三部分　基本实验操作与培养技术

实验十六　培养基的配制与灭菌

【实验目的】

1. 掌握培养基的配制方法。
2. 熟悉高压蒸汽灭菌的原理与操作。

【实验原理】

微生物的生长依赖于可利用的营养物质和适宜的生长环境。实验室条件下研究微生物需要供微生物生长的培养基。培养基是用于培养、转移、贮存微生物的固体、半固体或液体营养基质。培养基中必须含有微生物生长所需的全部营养物质，包括碳源、氮源、能源（有时可能与碳源相同）、无机盐、水分和生长因子。适宜的 pH 对微生物的生长也是必不可少的。

所用培养基的物理状态有三种：液体、半固体和固体。这些培养基的主要区别是在固体和半固体培养基中加有固化剂——琼脂。琼脂是从某种海藻中得到的结构复杂的长链多糖，具有很多优良性质。当加入溶液中时，琼脂要在 98℃ 才能融化形成有一定黏性的液体，并在 42℃ 凝固。凝固之后要重新加热至 98℃ 才能再次融化。琼脂具有的其他特性包括：不易被微生物降解；接近无色，因此易于观察菌落等。对于固体培养基需要加入琼脂的量约为 1.5%，半固体培养基需加约 0.5%。

图 16 – 1　高压蒸汽灭菌锅

Fig. 16 – 1 Auto clave

根据培养基成分的来源不同，可将培养基分为两种：一种叫限定培养基或合成培养基，由化学成分完全了解的物质配制而成，只能满足特定微生物的基本生长需求；另一种叫天然培养基，由化学成分不清楚或不恒定的物质组成，例如肉的水解产物（蛋白胨），能够充分满足不同微生物的生长需要。天然培养基如营养肉汤培养基在实验室中通常用于培养细菌。

培养基在配制过程中会由于接触到容器、药品、称量纸或其他表面而被污染，因此必须进行灭菌，同时不能将培养基中的营养物质破坏。这就需要高压蒸汽灭菌锅（图 16 – 1）在 121℃（即 1.05kg/cm^2 或 103.46kPa）条件下灭菌 15 ~ 30 分钟。在此条件下，所有微生物，包括芽孢在内，都会被杀死。当培养基中含有糖分的时候，灭菌温度也可降至 113℃。

【实验材料】

1. 试剂

牛肉浸膏、蛋白胨、氯化钠、琼脂、蒸馏水、1 mol/L HCl、1 mol/L NaOH。

2. 仪器

天平、药匙、称量纸、量筒、烧杯、pH 试纸、三角瓶、硫酸纸、高压蒸汽灭菌锅、电炉。

【实验方法】

1. 营养肉汤培养基的配制

配方如下：

牛肉膏	0.5g
蛋白胨	1.0g
氯化钠	0.5g
蒸馏水	100ml
pH	7.2～7.6

（1）按照配方准确称量每一种药品。牛肉膏黏性高，需用硫酸纸称取。

（2）用量筒称取 100ml 蒸馏水倒入烧杯。然后将各药品加入烧杯，加热并搅拌至溶解（将称好的牛肉膏与硫酸纸一并放入烧杯中，待溶解后将硫酸纸取出）。

（3）用 1 mol/L HCl 或 1 mol/L NaOH 调节 pH 至 7.2～7.6。

（4）将配好的培养基分装到三角瓶（试管）中至不超过 1/2 高度处。如果配制固体或半固体培养基，将称好的琼脂直接放入三角瓶（试管）中。盖好胶塞。

（5）灭菌：用牛皮纸包装好三角瓶（试管）并写上名字。然后将三角瓶（试管）放入高压蒸汽灭菌锅中灭菌 30 分钟。

（6）如果不立即使用，应将灭菌后的培养基置 4℃保存。

2. 高压蒸汽灭菌

（1）向锅内加入足够的水。

（2）放入配好的培养基，注意不要装得太挤。

（3）关闭锅盖，按下开始按钮。

（4）打开排气阀，待冷空气完全排尽后，关上排气阀。

（5）蒸汽由夹套进入内层。于是温度开始上升，当达到 121℃时，计时器开始计时 15～30 分钟（视所灭菌的培养基种类而定）。灭菌后，锅内的液体缓慢冷却后方可打开锅盖，以防止由于压力骤降引起液体喷溅。

（6）将培养基放入 37℃恒温箱培养 24 小时，检查是否灭菌彻底。

【结果与讨论】

1. 在配制培养细菌的培养基时为什么要将 pH 调至中性？
2. 灭菌结束后，为什么要等到灭菌锅内压力降为常压时才能打开灭菌锅盖？

PART THREE Basic Laboratory and Culture Techniques

16 Culture Media Preparation and Sterilization

Purposes

1. To be familiar with the preparation methods of culture media.
2. To know the Principles and the operation procedure of autoclave sterilization.

Principles

The growth of microorganisms depends on available nutrients and a favorable growth environment. Microbial laboratory works require the culture media to grow the microorganisms. A culture medium is a solid, semisolid or liquid preparation used to grow, transport and store microorganisms. The medium must contain all the nutrients the microorganisms require for growth, including carbon source, nitrogen source, energy source (can be the same as carbon source), inorganic source, water and growth factors. A suitable pH is also needed for the microbial growth.

Three physical states media are used: liquid media; semisolid media; and solid media. The major difference among these media is that solid and semisolid media contain a solidifying agent agar. Agar is a complex, long chain polysaccharide derived from certain marine algae and has several useful properties. When added to a solution, it melts at 98℃ forming a slightly viscous liquid that solidifies at 42℃. After solidification, the agar will not melt unless the temperature is again raised to 98℃. Some other useful properties of agar include its resistance to microbial degradation and its translucence for easy viewing of colonies. When a solid medium is necessary, a 1.5% concentration of agar is typical. For semi - solid medium around 0.5% agar is employed.

According to the sources of media ingredients, two different types of media are designed. A defined or synthetic medium contains known amounts of pure chemicals which are usually the basic requirements for the growth of particular microbes. A complex medium contains undefined ingredients such as protidtemns of meat (peptones) which provide enough nutrients to sustain the growth for different microbes. Complex media such as nutrient broth is routinely used for cultivation of bacteria in the laboratory works.

Culture media must be made sterile without inactivating nutrients necessary for growth of the microorganisms due to the contamination of containers, media ingredients, weighing papers, or other surfaces that come in contact with the medium. This involves the use of an autoclave (Fig. 16 – 1), in which media are sterilized by steam at 121℃ (1.05kg/cm^2 or 103.46kPa) for 15 ~ 30 minutes. Under these conditions, microbes, even endospores, will all be killed. The sterilization temperature could be decreased to 113℃ when sugar ingredients are involved.

Materials and Apparatus

1. Reagents

Beef extract, peptone, sodium chloride, agar, distilled water, 1 mol/L hydrochloric acid, 1 mol/L sodium hydroxide.

2. Apparatus

Balance, medicine spoon, weighing paper, cylinder, beaker, pH indicator paper, flask, vegetable parchment, autoclave, electric stove

Procedures

1. Preparation of nutrient broth

Recipe of nutrient broth is as follows:

Beef extract	0.5g
Peptone	1.0g
Sodium chloride	0.5g
Distilled water	100ml
pH	7.2 ~ 7.6

(1) Use the recipe listed above to weigh out each component required for the nutrient broth medium. For the beef extract, weigh it on a vegetable parchment because of its viscous property.

(2) Measure 100ml water using cylinder and then add all components into the beaker. Heat and stir until dissolved. (Add beef extract together with vegetable parchment into the beaker. After dissolving, take out the vegetable parchment.)

(3) Adjust the solution pH with 1mol/L HCl or 1mol/L NaOH to 7.2 ~ 7.6.

(4) Distribute the medium to flask (tube) less than half height of them. When solid or semisolid medium is needed, add agar into the flask (tube) directly. Put on

the plug.

(5) Sterilization: Wrap the flask (tube) with brown paper and write your name on it. Now the flask (tube) can be put into the autoclave for sterilization for 30 min.

(6) After sterilization, preserve the sterilized media at 4℃ if they are not used right now.

2. Sterilization

(1) Add enough water into the autoclave.

(2) Put the media into the chamber. Note: do not pack too tightly.

(3) Lock the door of the autoclave. Press the start button to begin sterilization

(4) Open the exhaust valves until the cool air is completely exhausted then close it.

(5) Steam enters into a jacket surrounding the chamber. The temperature in the chamber will go up to 121℃: at this point the timer begins to count down—15 to 30 min depending on the type of media. After sterilization, autoclaved liquids must be cooled slowly to avoid boiling over when the pressure is released.

(6) Place the media in the incubator at 37℃ for 24h to examine if they are sterilized completely.

Results and Discussion

1. Why are media for culturing bacteria adjusted to a neutral pH?

2. Why should the autoclave door be not opened after sterilization until the pressure in it lowers down to normal pressure?

实验十七　接种技术与微生物的分离纯化

【实验目的】

1. 理解纯培养物制备的原理。
2. 正确使用无菌操作进行接种。
3. 学会判断细菌在不同培养基上的生长状况。

【实验原理】

通常情况下，微生物以杂居状态生长。因此要对某一种微生物进行研究，必须先将其与其他微生物进行分离。纯培养物（即由单一微生物生长得到）是进行微生物实验操作的重要材料。微生物最容易的分离方法就是进行划线分离。

当通过划线法分离单菌落时，培养物在固体培养基表面通过划线进行稀释。目的是要使细菌细胞间的间隔足够大，这样单个细胞所产生的大量后代就会形成一个菌落。由此可见，菌落就是在固体培养基表面由微生物形成的肉眼可见的细胞集团，一个菌落就代表了一个纯培养物。

纯培养物也可以由涂布法获得。将少量微生物混合样本的稀释液加入到固体琼脂平板表面，并利用无菌涂布棒涂布均匀。如果稀释度合适，分散的细胞就会形成单菌落。

上面的分离步骤都涉及到接种技术。接种所使用的器材包括接种针、接种环和移液管。接种针（环）用于将微生物从液体或斜面培养物中转移至其他培养基。接种针和接种环都由把手、金属柄、转头以及固定的镍镉或铝金属丝组成。移液管则是用来转移液体培养物，或用于微生物的梯度稀释以及配置化学药品。

分离纯培养物时，必须使用无菌操作来防止由于容器、药品、仪器和其他表面接触培养物所带来的污染。最重要的分离和接种工具就是接种针（环）。接种针（环）可以通过使用煤气灯火焰加热至红色来快速灭菌。

当对未知菌种进行鉴定时，首先要制备出纯培养物，然后通过微生物的培养特征、细胞特征和生化特征进行分析。其中最简单的是培养特征。在固体培养基上，分离出的单菌落可以通过大小、形状、边缘、色素、质地和表面等进行观察。在液体培养基中，三种典型的生长形式包括均匀混浊生长、只在液体表面生长以及只在管底生长。半固体培养基则用于检验细菌的运动性。当在半固体培养基中央进行穿刺时，在穿刺线周围的生长状况就会显示样本的运动性（图 17－1）。如果培养基透明，只沿穿刺线生长（左），说明菌体不具有运动性；当培养基显示云雾状时（右），则说明菌体具有运动性。

图 17－1　半固体培养基中的生长状况

Fig. 17－1 Growth patterns in semisolid media

【实验材料】

1. 菌种

20 小时金黄色葡萄球菌斜面培养物，20 小时大肠埃希菌斜面培养物，20 小时藤黄八叠球菌斜面培养物，24 小时枯草芽孢杆菌斜面培养物，20 小时混合的细菌液体培养物（金黄色葡萄球菌与大肠埃希菌）。

2. 培养基

营养肉汤琼脂斜面，营养肉汤培养基，半固体培养基，琼脂平板。

3. 仪器

无菌平皿，接种环，接种针，无菌移液管，煤气灯。

【实验方法】

1. 接种技术（图 17－2）

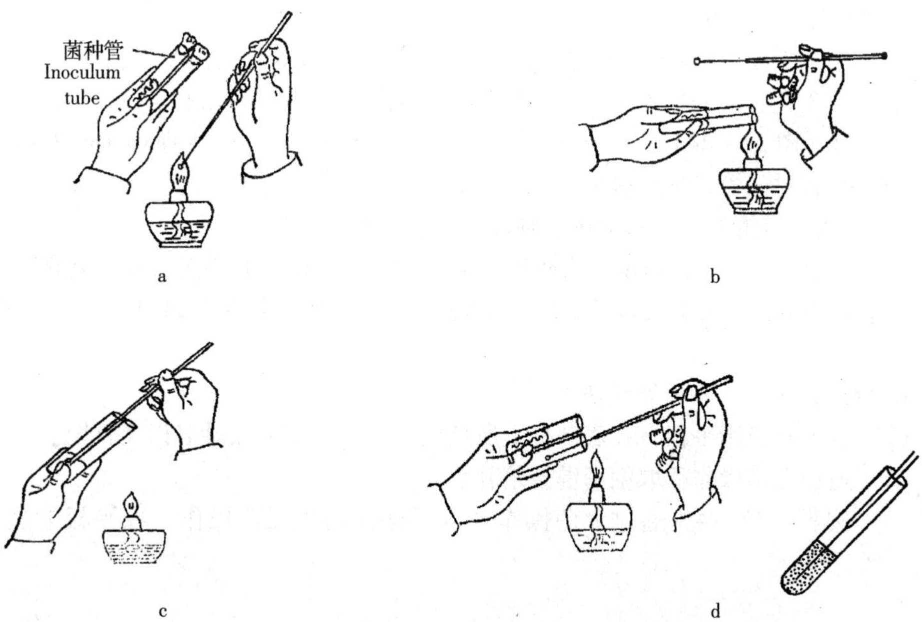

图 17-2　无菌接种法

Fig. 17-2 Aseptic techniques for bacterial removal and subculturing

(1) 将要接种的试管或平皿写好名字、日期、所用培养基和微生物名称。

(2) 调整火焰至适当大小（蓝色）。如果原始培养物是液态，先要将其摇晃均匀。用左手握住原培养物试管与待接种试管。右手持接种环（针）（图 17-2a）。

(3) 用火焰加热接种环（针）至变红，然后沿金属柄向上，将能深入试管的部分依次灭菌（图 17-3）。

(4) 待接种环（针）冷却后再取菌。否则细胞会被高温杀死。

(5) 再次确认所取微生物样本。

(6) 右手打开试管帽，于手掌与小指间夹住。在打开或盖上试管帽时都要用火焰对试管口进行灭菌。（图 17-2b）

(7) 用灭菌的接种环从原培养物试管中取一环培养物（图 17-2c）加入到新的液体培养基中（图 17-2d）。在这一步，也可以转移至载玻片或在斜面上划线，还可以从原始斜面培养物进行接种。

(8) 对于半固体穿刺接种，将接种针插入装有半固体培养基的试管中至 2/3 处（注意不要碰到试管壁）然后沿穿刺线拔出（图 17-2d）。

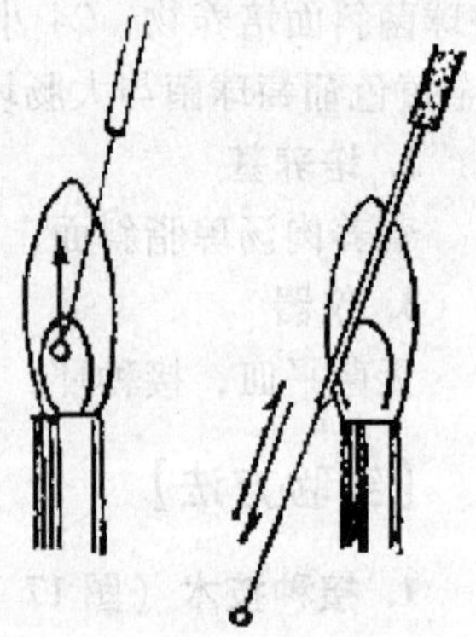

图 17-3　接种环（针）的灭菌

Fig. 17-3 Sterilization of inoculating loop (needle)

(9) 重新对试管口进行灭菌（图 17-2e）。

(10) 盖上试管帽，再次灭菌接种环（针）（图 17-2f）。

分别接种金黄色葡萄球菌、大肠埃希菌和枯草芽孢杆菌的斜面培养物至液体培养基和半固体培养基中，将所有接种后的培养物放入 37℃恒温箱中培养 24 小时，然后观察培养结果。

2. 分离纯化——平板划线法

(1) 准备好琼脂平板，在底部标上所接菌种名称、实验人姓名以及日期。

(2) 通过无菌操作移取细菌混合培养物。

(3) 按照图 17-4，通过无菌操作（以下操作均为无菌操作）在平板上进行划线，得到一区。

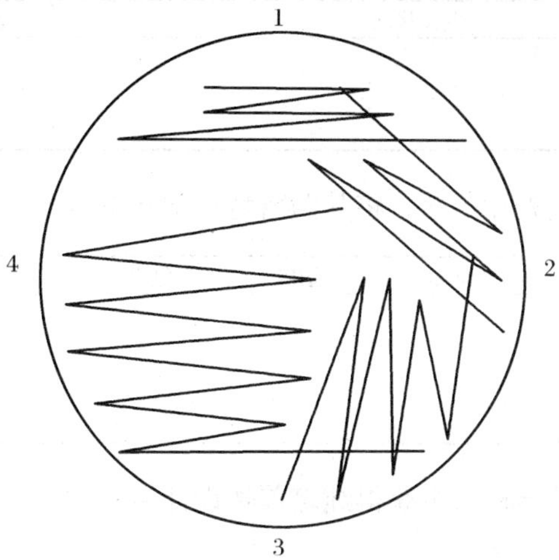

图 17 – 4　四区划线

Fig. 17 – 4 Quadrant streaking

1. 一区生长最为茂盛；
2. 二区由一区移种，其生长密度低于一区；
3. 三区由二区移种，因此菌体数量小于二区；
4. 四区由三区移种，菌数最少，出现单菌落。

（4）将接种环移开并灭菌，杀死残留在上面的菌体。待在非接种区冷却后，从一区出发，划出二区。

（5）再次灭菌接种环，重复上一步骤，得到三区。

（6）同法得到四区。

（7）将划线后的平板倒置在 37℃ 恒温箱中培养 24 小时，观察培养结果。

（8）按照相同的方法对枯草芽孢杆菌和藤黄八叠球菌的斜面培养物进行分离纯化。

3. 分离纯化——涂布法

（1）准备好琼脂平板，在底部标上所接菌种名称、实验人姓名以及日期。

（2）用无菌滴管移取 0. 1ml 细菌混合培养物至琼脂平板表面。

（3）在无菌区，用无菌涂布棒将混菌样本涂布均匀并覆盖整个平板表面。

（4）将用过的涂布棒置于指定位置待灭菌。

（5）将平板放置在 37℃ 恒温箱中倒置培养 24 小时，观察培养结果。

【结果与讨论】

1. 观察琼脂斜面上各细菌的生长状况。
2. 观察液体培养基中细菌生长状况，不得用力振荡培养物，将实验结果记于下表中。

菌　名	生长状况
金黄色葡萄球菌	
大肠埃希菌	
枯草芽孢杆菌	

3. 观察并记录细菌在半固体培养基中生长状况，结果记于下表。

菌　名	生长状况	运动性
金黄色葡萄球菌		
枯草芽孢杆菌		
大肠埃希菌		

4. 观察不同细菌的代表性菌落，将其菌落特点记于下表。

菌　名	大小	形状	边缘	色素	质地	表面
金黄色葡萄球菌						
大肠埃希菌						
枯草芽孢杆菌						
藤黄八叠球菌						

5. 为从一个微生物混合材料中获得纯培养物设计两种不同的实验程序。

17 Culture Transfer Techniques and Isolation of Pure Cultures

Purposes

1. To understand the Principles behind pure culture preparation.

2. To know how to correctly use aseptic technique to transfer bacteria for subculturing

3. To know how to determine the growth characteristics of bacteria on different media.

Principles

Commonly, microorganisms usually grow in complex and mixed population containing many different species. To get adequate study, single type of microbial sample has to be separated from other species. Pure culture, the growth of a single species of microorganism, is of great importance in microbial laboratory techniques. The easiest and best way to do this is to streak for isolation.

When streaking for isolated colonies, the inoculum is diluted by streaking it across the agar plate surface. The purpose is to get the inoculum diluted to the point where there is only one bacterial cell deposited every few millimeters on the surface of the agar plate. When these lone bacterial cells divide and give rise to thousands of offspring, an isolated colony is formed. Therefore, a colony is a macroscopically visible growth or cluster of microorganisms on a solid medium and each colony represents a pure culture.

Pure culture can also be obtained from spread plate. A small volume of dilute microbial mixture is transferred to the surface of an agar plate and spread with a sterile glass spreader. If the dilution is proper, the dispersed cells develop into isolated colonies.

All the above isolation procedures involve culture transfer techniques. Culture transfer instruments include inoculating needles, inoculating loops and pipette. Inoculating needles and loops are used to aseptically transfer microorganisms from broth or slant cultures to other media. Both may consist of handles, a shaft, and a turret, which holds a nickel – chromium or platinum wire. A pipette is often used to transfer aliquots of culture, to prepare serial dilutions of microorganisms, and to dispense chemical reagents.

To isolate pure culture, aseptic techniques must be used to avoid contamination of containers, media ingredients, instruments and other surfaces that come in contact with the culture. The most important tool for isolation and transferring cultures is the wire inoculating needle or loop. It can be quickly sterilized by heating it to red hot in a Bunsen burner flame.

If an unknown bacterium sample needs to be identified, pure culture should firstly prepare. Then different microbial characteristics including cultural characteristics, cellular characteristics, and biochemical characteristics could be examined. The simplest of these to observe are the cultural characteristics. On the agar plate (solid media), well - isolated colonies are evaluated according to the following parameters: size, shape, margin, pigment, texture and surface. In nutrient broth (liquid media), three typical growth patterns include growth turbid and diffuse throughout, growth layered at surface only and growth sedimented at bottom only. Semisolid media are used to determine the motility of bacteria. When the medium is stabbed carefully in the center, the growth pattern around the stab line shows the motility of the sample (fig. 17 - 1). If the media is clear with a persistent stab line, the organism is non - motile. If the media gives a uniform cloudiness, then the organism is motile.

Materials and Apparatus

1. Cultures

Nutrient agar slants of *Staphylococcus aureus*, *Escherichia coli*, *Bacillus subtilis* and *Sarcina lutea*, mixed broth culture of *S. aureus* and *E. coli.*

2. Media

Nutrient agar slants, nutrient broth, semisolid media, agar plates.

3. Apparatus

Sterile plate, inoculation loop and needle, sterile pipette, burner.

Procedures

1. Culture transfer techniques (shown in fig. 17 - 2)

(1) Label the tubes and/or plates you will be inoculating with your name, the date, the type of medium it is and the name of the organisms.

(2) Properly adjust the flame of the burner. The proper flame is a small blue cone not a large plume. If the primary culture is broth, mix the tube to get a uniform suspension. Place the stock culture tube and the tube to be inoculated in the palm of the left hand (Fig. 17 - 2a).

(3) Flame your loop (or needle) until the wire becomes red – hot. Then sterilize the metal handle which will be inserted into the tube (Fig. 17 – 3) .

(4) Allow your loop (or needle) to cool before you try to pick up your organism. Otherwise, the cells will be killed by heat.

(5) Ensure that you are transferring the correct organism into your labeled tube by double – checking the name of the organism on the stock culture from which you are collecting your inoculum.

(6) When removing the caps from tubes, always keep the caps between the palm and the small fingers. (Fig. 17 – 2b) And also flame the lip of a culture tube when you open it and before you replace the cap.

(7) With the sterile inoculating loop, transfer 1 loopful of culture from the stock culture tube (Fig. 17 – 2c) into the new broth tube (Fig. 17 – 2d) . At this point, one could also transfer to a glass slide or streak the surface of a slant. The original culture could be a slant culture as well.

(8) For the semisolid stab transfer technique, move the inoculating needle into the semisolid media tube without touching the walls of the tube, and the needle penetrates medium 2/3 depth. Then take the needle out along the stab line (Fig. 17 – 2d) .

(9) Reflame the neck of the tubes (Fig. 17 – 2e) .

(10) Recap the tubes and sterilize the loop (or needle) (Fig. 17 – 2f) .

Inoculate *S. aureus*, *E. coli* and *B. subtilis* from the slant cultures into nutrient broth and semi – solid media respectively. Incubate at 37℃ for 24h. Then observe the results.

2. Isolation of pure culture—streaking – plate technique

(1) Prepare an agar plate and mark on the bottom of the plate with the name of the bacteria to be inoculated, your name, and date.

(2) Aseptically remove a loopful of the bacterial mixture.

(3) Using Fig. 17 – 4, streak the first section of your plate. This has to be done within the aseptic area. (same as the following steps)

(4) Remove the inoculating loop and kill any remaining bacteria by flaming them. Cool the loop at an uninoculated section of the agar plate. Cross over the streaks in area 1 and streak out as shown in area 2.

(5) Flame your loop again, and repeat the streaking process for your third section.

(6) Repeat this sequence once more to make area 4.

(7) Incubate the plates upside down at 37℃ for 24h. Then observe the isolation results.

(8) Do the same to purify *B. subtilis* and *S. lutea* from their slant cultures.

3. **Isolation of pure culture—spread - plate technique**

(1) Prepare an agar plate and mark on the bottom of the plate with the name of the bacteria to be inoculated, your name, and date.

(2) Make a series of dilution tubes of the original mixed inoculum. Each succeeding dilution tube has one - tenth the number of microbial cells as the preceding one.

(3) For each dilution tube, aseptically pipette 0.1 ml of the bacterial mixture onto the center of an agar plate.

(4) In the aseptic area, use the sterile glass spreader to spread the bacterial sample evenly over the agar surface, making sure that the whole surface has been covered.

(5) Put the spreader in a proper container to sterilize.

(6) Incubate the plates upside down at 37℃ for 24h. Then observe the isolation results.

Results and Discussion

1. Observe the cultures on agar slants.

2. Observe the cultures in nutrient broth without shaking and write down the cultural characteristics of different bacteria.

Strains	Cultural characteristics
S. aureus	
E. coli	
B. subtilis	

3. Observe the growth patterns of cultures in semisolid media and write down the results in the following table.

Strain	Cultural characteristics	Motility
S. aureus		
B. subtilis		
E. coli		

4. Observe the characteristics of the representative colonies of different bacteria and write down the results in the following table.

Strains	Size	Shape	Margin	Pigment	Texture	Surface
S. aureus						
E. coli						
B. subtilis						
S. lutea						

5. Design two different practical methods to obtain pure culture from mixed sample.

实验十八　常用的消毒灭菌方法

【实验目的】

1. 熟悉巴氏消毒、紫外杀菌和干热灭菌的原理和步骤。
2. 验证细菌芽孢的耐热性。

【实验原理】

除高压蒸汽灭菌外，对于不同物体还有其他消毒灭菌方法。本实验将介绍巴氏消毒、紫外杀菌和干热灭菌。每一种方法的使用特点与限制将通过杀菌效果展示出来。

巴氏消毒是用于对牛奶或其他液体进行加热以破坏腐败或病原微生物的消毒措施。操作温度通常在63～72℃，因此只能杀死营养细胞，对芽孢无效。

紫外线的波长为100～400nm。当微生物暴露在紫外光下时，DNA会吸收紫外线，并使其中相邻的胸腺嘧啶相连成二聚体。DNA复制时，无法对此二聚体进行配对，引起DNA链复制错误。换句话说，紫外照射能够引起突变并导致蛋白质合成发生错误。如果突变过多，细菌的代谢就会被阻断，从而引起微生物死亡。由此可知，紫外照射不能用于杀灭产芽孢的细菌，因为处于芽孢这一休眠时期，没有DNA复制发生。

紫外线的杀菌能力受照射时间长短的影响，照射时间越长，杀菌效果越好。同时也由所用紫外线波长决定。260nm的紫外线具有最强杀菌性，因为DNA分子的最大吸收峰在此。由于紫外照射穿透力差，只能用于表面杀菌。紫外线对眼睛和皮肤的细胞会造成伤害。

许多物品需要在干燥条件下进行灭菌，即干热灭菌。将待灭菌物品置于烘箱中160～170℃灭菌1～2小时，可以保证彻底杀灭包括芽孢在内的一切微生物。这是因为高温能够使蛋白质氧化变性。适于干热灭菌的物品有：玻璃仪器、金属仪器和粉末等水蒸气无法有效穿入并且在高温下不会被破坏的物品。

【实验材料】

1. 菌种

24小时枯草芽孢杆菌斜面和液体培养物，20小时大肠埃希菌斜面和液体培养物。

2. 仪器

接种环，温控水浴锅，琼脂平板，营养肉汤培养基试管，烘箱，紫外灯，培养箱，无菌滴管等。

【实验步骤】

1. 巴氏消毒

（1）分别接种枯草芽孢杆菌和大肠埃希菌到试管内的营养肉汤培养基中，再用另一管未接种的培养基作为空白对照。

（2）将3个试管共同置于63℃恒温水浴30分钟进行消毒。

（3）将试管置于37℃恒温培养24小时，观察细菌在消毒后的生长状况。

2. 紫外杀菌

（1）打开紫外灯预热10分钟。

（2）用无菌滴管分别滴加3滴两种液体培养物至两个琼脂平板表面，并涂布均匀（需无菌操作）。

（3）用灭菌后的镊子取一张无菌的星形牛皮纸（需无菌操作）。

（4）将牛皮纸放至平板中央并将边缘按压（需无菌操作）。

（5）移去平板盖，在距紫外灯20cm左右距离照射30分钟。注意不要将皮肤或眼睛直接暴露在紫外灯下，以免受伤。

（6）照射后，用镊子通过无菌操作将牛皮纸取下，重新盖好，在37℃培养24小时。

（7）培养后，观察微生物的生长情况。

3. 干热灭菌

（1）将待灭菌物品（平皿、试管、吸管等）放入电烘箱内，注意物品不要摆得太挤。

（2）打开烘箱开关，直至达到160～170℃，开始计时2小时。

（3）待电烘箱内温度降到70℃以下取出灭菌物品。

【结果与讨论】

1. 分别记录并比较63℃水浴和紫外照射对两种培养物的消毒灭菌效果。
2. 干热灭菌的注意事项有哪些？
3. 怎样才能获得枯草芽孢杆菌的芽孢悬液？

18 Commonly Used Sterilization and Disinfection Methods

Purposes

1. To be familiar with the Principles and procedures of Pasteurization, UV radiation and dry heat sterilization.

2. To verify the heat – resistant ability of endospores.

Principles

Besides autoclaving, there are some other sterilization methods for different objects. This exercise includes Pasteurization, UV radiation and dry heat sterilization. According to the bactericidal effects, the characteristics and limitations of each method will be shown.

Pasteurization is the process of heating milk and other liquids to destroy microorganisms that can cause spoilage or disease. The manipulation temperature is usually between 63 to 72 ℃. Therefore, only vegetative cells can be destroyed not the endospore.

Ultraviolet ranges in wavelength from around 100nm to 400nm. When the microorganisms are exposed to UV, it is absorbed by microbial DNA and causes adjacent thymine bases to covalently bond together, forming TT dimers. As the DNA replicates, nucleotides do not complementary base pair with the TT dimers and this results in mistakes in the replication of that DNA strand. In other words, UV radiation causes mutation and can lead to faulty protein synthesis. With sufficient mutation, bacterial metabolism is blocked and the organism dies. Based on this principle, UV radiation cannot be used to kill endospore – forming bacteria since no DNA replication is progressing in this dormancy stage.

The microbicidal activity of ultraviolet light depends on the length of exposure: the longer the exposure the greater the microbicidal activity. It also depends on the wavelength of UV used. The most lethal wavelength of UV light is 260 nm where the highest absorption peak is. UV radiation can only be used for surface sterilization because of its poor penetration power. It can also damage eyes, cause burns, and cause mutation in cells of the skin.

Many objects are best sterilized in the absence of water by dry heat sterilization even though it is not as versatile as autoclaving. The objects are placed in an over at 160 to 170℃ for 1 to 2h, which ensures thorough heating of the items and destruction of endospres. Microbial death apparently results from the oxidation and denaturation of proteins. Objects suitable for dry heat sterilization are glassware, metal instruments and powders that steam dose not penetrate well and cannot be destroyed at high temperature.

Materials and Apparatus

1. Cultures

24h slant and broth cultures of *Bacillus subtilis* and 20h slant and broth cultures of *Escherichia coli.*

2. Apparatus

Inoculating loop, temperature – controlled water bath, sterile agar plates and nutrient broth in tubes, oven, UV lamp, incubator, sterile dropper.

Procedures

1. Pasteurization

(1) Inoculate *B. subtilis* and *E. coli* to the nutrient broth tubes respectively and take another broth tube without inoculation as contrast.

(2) Place the three tubes together in the 63℃ water bath for 30 min for disinfection.

(3) After that incubate the tubes at 37℃ for 24 hours then record the disinfection results.

2. Ultraviolet radiation

(1) Switch on the UV lamp and allow it to warm up for about 10 min.

(2) Using sterile droppers add 3 drops of the two broth cultures onto two agar plate surfaces respectively and spread uniformly to make a lawn. All these need to be done with aseptic techniques including the following two steps.

(3) Sterilize the forceps and pick up a sterile brown paper in star shape.

(4) Put the brown paper on the center of the agar plate and press the edge.

(5) Remove the lid and expose the agar plate to the UV lamp at a distance of 20 cm for 30min. ALWAYS remind yourself not to expose your skin or eyes to the UV light directly.

(6) After exposure, take out the brown paper with sterilized forceps aseptically and recover the agar plate. Incubate at 37℃ for 24h.

(7) After incubation, observe the growth condition of the bacteria samples.

3. Dry heat sterilization

(1) Place the items (plates, tubes, straws, *et al*) in the oven. Note: Do not pack too much.

(2) Switch on the oven and wait until the inside temperature increases to 160 ~ 170℃. Time for 2h.

(3) Take out the items after sterilization (must wait the temperature to decrease to 70℃).

Results and Discussions

1. Respectively record and compare the sterilization results of the two cultures with 63℃ water bath and UV radiation.

2. What should we pay attention to in dry heat sterilization?

3. How to obtain spore suspension of *B. subtilis*?

实验十九　菌种保藏

【实验目的】

掌握菌种保藏的目的、基本原理及几种菌种保藏的常规方法。

【基本原理】

微生物具有容易变异的特性，因此，在保藏过程中，必须使微生物的代谢处于最不活跃或相对静止的状态，才能在一定的时间内使其不发生变异而又保持生命活动能力。

菌种保藏的目的是为了把从自然界分离到的野生型、或者经过人工选育得到的变异型纯种，采用多种方法使菌种存活、不丢失、不污染杂菌、不发生或少发生变异，保持菌种原有的各种优良培养特征和生理活性，有利于生产、科研的正常进行，是一项重要的微生物学基础工作。

微生物菌种保藏的基本原理是使微生物的生命活动处于半永久性的休眠状态，也就是使微生物的新陈代谢作用限制在最低的范围内。干燥、低温和隔绝空气是使微生物代谢能力降低的重要因素，是保证获得这种状态的主要措施。有针对性地创造干燥、低温和隔绝空气的外界条件，是微生物菌种保藏的基本技术。尽管菌种保藏方法很多，但基本都是根据这三种主要因素设计的。

菌种保藏方法大致可分为以下几种。

1. 传代培养保藏法

这是最简单的保藏方法。对需氧菌可用斜面培养，对厌氧菌可进行穿刺培养或疱肉培养基培养等，培养后于4℃冰箱内保存。此法为实验室和工厂菌种室常用的保藏法，优点是操作简单、使用方便、不需特殊设备；缺点是传代次数多，容易变异、污染杂菌。若菌种经常使用，可应用此法。

保藏时间依微生物的种类而不同。霉菌、放线菌及有芽孢的细菌可保存2~4个月，移种一次；酵母菌可保存2个月；细菌最好每月移种一次。

2. 液体石蜡覆盖保藏法

它是在斜面培养物和穿刺培养物上面覆盖1cm灭菌的液体石蜡后于4℃直立放置保存的方法。这样可防止固体培养基的水分蒸发而引起的菌种死亡，另一方面液体石蜡可阻止氧气进入，使好氧菌不能继续生长，从而延长了菌种保藏的时间，此法实用效果好。霉菌、放线菌、芽孢菌可保藏2年以上不死，酵母菌可保藏1~2年，一般无芽孢细菌也可保存1年左右，甚至用一般方法很难保藏的脑膜炎球菌，在37℃恒温箱内，亦可保藏3个月之久。此法的优点是操作简单，不需特殊设备；缺点是在保存过程中，菌种必须直立放置，所占空间较大。

3. 沙土保藏法

去除水分可以降低微生物的代谢速度，因而这一方法对可形成孢子的微生物比如放线菌和真菌更加有效。因此在抗生素生产中应用最广、效果最好。操作是将微生物吸附在适当的载体，如土壤、沙子上，而后进行干燥。可保存2～10年左右，但应用于营养细胞效果不佳。

4. 超低温冷冻保藏法

可分低温冰箱（-20～-30℃，-50～-80℃）、干冰酒精快速冻结（约-70℃）和液氮（-196℃）保藏法等。尤其液氮冷冻保藏法，除适用于一般微生物的保藏外，对支原体、衣原体、氢细菌、难以形成孢子的霉菌、噬菌体及动物细胞等一些用冷冻干燥法难以保存的微生物，均可用此法长期保藏。其优点是性状不变异，但需特殊设备。

5. 冷冻干燥保藏法

先使微生物在低温度（-70℃左右）下快速冷冻，然后在减压下利用升华现象除去水分（真空干燥），利用有利于菌种保藏的一切因素，使微生物始终处于低温、干燥、缺氧的条件下，因而它是迄今为止最有效的菌种保藏法之一。

用冷冻干燥保藏的菌种，其保藏期可达数年至数十年，其菌种要特别注意纯度，不能污染杂菌，这样再次使用该菌种时才不会出差错。细菌和酵母菌菌种要求培养到稳定期，一般细菌培养24～48小时，酵母菌培养3天，放线菌与霉菌一般需培养7～10天，若用对数生长期菌种进行保藏，其存活率反而会降低。此方法保存孢子比其营养体效果更好。

后两种方法在冷冻过程中进行，为了避免恶劣条件对微生物的损害，需使用保护剂来制备细胞悬液，可以防止因冷冻或水分不断升华对细胞的损害。常用作保护剂的有脱脂牛奶、血清、糖类、甘油和二甲基亚砜等。

对于目前尚不能在人工培养基上生长的微生物，如病毒、立克次体、螺旋体等，可用寄主保藏法，即将它们在活的动物、昆虫、鸡胚内感染并传代，此法相当于一般微生物的传代培养保藏法。

【实验材料】

1. 菌种

细菌、酵母菌，放线菌和霉菌。

2. 培养基及其他材料

固体斜面培养基，灭菌脱脂牛乳，灭菌水，液体石蜡，五氧化二磷，河沙，黄土，冰块，食盐，干冰，95%乙醇，10%盐酸，无水氯化钙。

3. 仪器及其他

灭菌吸管，灭菌培养皿，安瓿，40目与100目筛子，油纸，干燥器，真空泵，喷灯，低温冰箱（-30℃），超低温冰箱，液氮冷冻保藏器。

【实验方法】

1. 传代培养保藏法

（1）斜面低温保藏法　将菌种接种于固体斜面培养基上，置适宜温度培养，待菌充分生长后，移至4℃的冰箱中保藏。

（2）穿刺法　将半固体培养基注入小试管（0.8cm×10cm），使培养基距离试管口约2～3cm深。用灭过菌的接菌针无菌操作挑取菌体，在半固体培养基顶部的中央直线穿刺到半固体培养基的1/3深处，在适宜条件下培养后，熔封试管或是塞上橡皮塞，移至4～6℃的冰箱中保藏。

2. 液体石蜡保藏法

（1）将液体石蜡高温灭菌（121.3℃，30分钟）。

（2）将菌种接种至斜面培养基中培养，使其充分生长。

（3）用无菌吸管吸取液体石蜡注入斜面，其用量以高出斜面顶端1cm为准。

（4）保持试管直立，置4℃保存。

从液体石蜡下面取培养物移种后，接种环在火焰烧灼时，培养物容易与残留的液体石蜡一起飞溅，应特别注意。

3. 沙土保藏法

（1）取河沙加入10%稀盐酸浸泡，去除有机杂质。

（2）倒去酸水，用自来水冲洗至中性，烘干。用40目筛子过筛，去掉粗颗粒，备用。

（3）取非耕作层的瘦黄土，磨细，100目筛子过筛。

（4）按一份土、三份沙掺合均匀，装入小试管中，每管装1cm左右，灭菌。

（5）抽样进行无菌检查，每10支沙土管抽一支，将沙土倒入肉汤培养基中，37℃培养48小时，若仍有杂菌，则需重新灭菌。

（6）接种方法分干法接种和湿法接种。

干法接种（多用于保藏放线菌和部分真菌）：将斜面上已生长好的培养物，用接种环或接种铲刮取孢子，接种于无菌的沙土管中，搅拌均匀，注意勿接入培养基。

湿法接种：选择培养成熟的（一般指孢子层生长丰满的）优良菌种，将已生长好的菌种斜面培养物用3～5ml无菌水洗下，制成均匀的菌悬液，再用无菌吸管吸取菌悬液加到沙管中，每管加10滴约0.5ml。接种后的沙土管放于真空泵中抽干以除去沙土管中的水分，抽干时间越短越好。

（7）每10支抽取一支，用接种环取出少数沙粒，接种于斜面上，培养后观察生长情况和有无杂菌。

（8）若经检查没有问题，用火焰熔封管口，放冰箱或室内干燥处保存。每半年检查一次活力和杂菌情况。

（9）菌种复活培养时，取沙土少许移入液体培养基内培养。

4. 液氮超低温冷冻保藏法

（1）准备安瓿管　用于液氮保藏的安瓿管，需要采用硼硅酸盐玻璃制成的，能耐受温度突然变化而不致破裂。安瓿管的大小通常为7.5 mm×100mm，能容1.2ml液体。将空安瓿管塞上棉塞，高压灭菌。

（2）准备保护剂与灭菌　保护剂如10%的甘油或10%二甲亚砜。含甘油溶液需经高压灭菌，含二甲亚砜溶液则采用过滤除菌。保存细菌、酵母菌或霉菌孢子等容易分散的细胞时，则将空安瓿管塞上棉塞；若保存霉菌菌丝体，则需在安瓿管内预先加入保护剂如10%二甲基亚砜蒸馏水溶液或10%的甘油蒸馏水溶液，加入量以能浸没以后加入的菌块为限。0.1MPa 高压蒸汽灭菌 20 分钟。

（3）接入菌种　将菌种用 10% 的甘油蒸馏水溶液制成菌悬液，装入无菌的安瓿管；霉菌菌丝体可用无菌打孔器，从平板内切取菌落圆块，放入装有保护剂的安瓿管内，然后用火焰熔封，浸入水中检查有无漏洞。

（4）冻结　将已封口的安瓿管以每分钟下降 1℃ 的慢速冻结至 -30℃。避免细胞急剧冷冻在细胞内，这样会形成冰的结晶，降低存活率。

（5）保藏　将冻结至 -30℃ 的安瓿管立即放入液氮冷冻保藏器的小圆筒内，再将小圆筒放入液氮保藏器内。液氮保藏器内的气相为 -150℃，液态氮内为 -196℃。

（6）恢复培养　需要用保藏的菌种时，将安瓿管取出，立即放入 38～40℃ 水浴急速解冻，直到全部溶化为止。再采用无菌操作技术，打开安瓿管，将内容物移入适宜的培养基上培养。

5. 真空冷冻干燥保藏法

（1）准备安瓿管　安瓿管可用长颈球形底的，材料为中性硬质玻璃。安瓿管先用 2% 盐酸液浸泡 8～10 小时，再用自来水冲洗多次，最后用蒸馏水洗 2～3 次后烘干。将印有菌名和接种日期的标签放入安瓿管内（有字的一面应朝向管壁），管口塞上棉花，0.1MPa 高压蒸汽灭菌 30 分钟。

（2）准备保护剂——脱脂牛奶　先将鲜牛奶煮沸，除去上面的一层脂肪，然后用脱脂棉过滤，并在 3000 转/分钟离心 15 分钟。如果一次不行，再离心一次，直至除尽脂肪为止。牛奶脱脂以后，在 55.2 kPa 条件下，高压蒸汽灭菌 30 分钟，并作无菌检验。

（3）准备保藏菌种　所用菌种要特别注意其纯度，不能有杂菌污染。用无菌吸管吸取 2～3ml 脱脂牛奶加到待保藏的菌种斜面内，用接种环将菌种刮下，轻轻搅动，使其均匀地悬浮在牛奶内成菌悬液（但应注意，切勿将琼脂刮到牛奶中），每支安瓿管分装 0.2ml（一般装入量为安瓿管球部体积的 1/3 为宜）。

（4）预冻　将分装好的安瓿管放于低温冰箱或干冰酒精液中（-25～-40℃之间）冷冻，使悬液结冰。

（5）真空干燥　将装有已冻结菌悬液的安瓿管置于真空干燥箱真空干燥，一般抽到真空度 0.0267～0.0133 MPa 维持 6～8 小时，样品可被干燥。

（6）封管　取出安瓿管，接在封口用的抽气装置上抽气，当真空度达到 0.0267MPa 时，继续抽气数分钟，用火焰在细颈处烧熔封口。做好的安瓿管应放置在低温避光处保藏（图 19-1）。

（7）恢复培养　如果要从中取出菌种恢复培养，可先用 75% 乙醇将管的外壁消毒，然后将安瓿管上部在火焰上烧灼，滴几滴无菌水，使管子破裂。再用接种针直接挑取松散的干燥样品，在斜面接种。也可先将无菌液体培养基加入安瓿管中，使样品溶解。然后再用无菌吸管取出菌液至合适的培养基中进行培养。

此法为菌种保藏最有效的方法之一，对一般活力强的微生物及其孢子以及无芽孢菌都适用，即使对一些很难保存的致病菌，如脑膜炎球菌与淋病球菌等亦能保存。

【结果与讨论】

1. 经常使用的细菌菌种，应用哪一种保藏方法既好又简便？

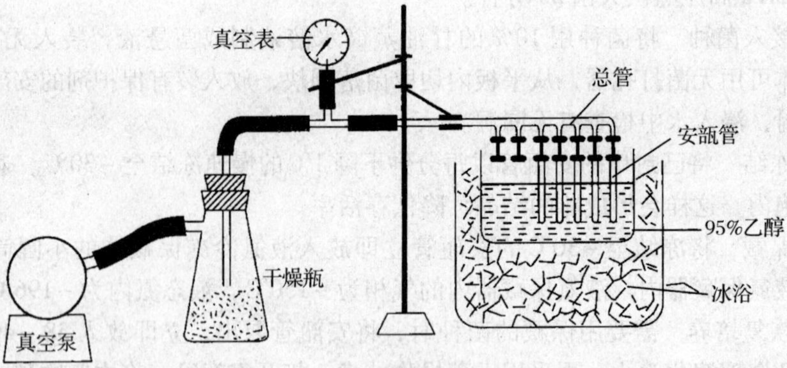

图 19－1　真空冷冻干燥保藏法

Fig. 19－1 The freeze－drying method

2. 细菌用什么方法保藏的时间长而又不易变异？
3. 产孢子的微生物常用哪一种方法保藏？

19 Preservation of Pure Cultures

Purpose

1. To understand the purpose and basic principles of pure culture preservation.
2. To be familiar with the routine methods of pure culture preservation.

Principle

Microorganisms have the characteristic of easy variation, so in the process of preservation, relative inactivity of microbial metabolism can prevent organisms from variation and permit the survival in a certain period.

The purpose of culture preservation is to maintain the wild cultures that isolated from the nature or the variable pure cultures obtained by artificial selection without loss, contamination or variation. And to keep the original features and physiological activities of the cultures and finally to facilitate the research are also involved. Therefore, these techniques are one of the foundation in microbiology.

The basic principle of culture preservation is to keep the microorganisms in a semi – perpetual dormancy stage, which can minimize the metabolism activity of microorganisms. The most important factors of decreasing the metabolism are desiccation, low temperature and isolation from air. Consequently, they are used as the main techniques for preserving cultures. Although there are many methods of culture maintenance, they are all designed according to the above three techniques.

The methods for culture preservation can be classified as follows:

1. Subculturing

It is the simplest method of preservation. For aerobes, they are cultivated on agar slants. While for anaerobes, they are cultivated by stab culture or cooked meat medium. Store the cultures at 4℃ for several weeks. The method is commonly used in the laboratory and factory. The advantage of which is easy to operate, convenient to use and no need of special equipment though the strain variation and contamination may occur due to high passage.

The period of preservation is different in species of microorganisms. For molds, actinomycetes and zygotomere, they can be preserved for 2 ~ 4 months. For yeasts, they can be preserved for 2 months. For bacteria, it is better to subculture once a month.

2. Covered with liquid paraffin

It is to cover the slant or stab cultures with 1cm of sterile liquid paraffin and then place them erectly at 4℃. For one thing it can prevent the death of cultures due to the evaporation of water in solid culture medium, for another, liquid paraffin can prevent the entry of oxygen which can suppress the growth of aerobes in order to prolong the time of preservation. This method is very practical and also has good effect. We can preserve molds, actinomycetes and spores for more than 2 years; yeasts for 1 ~ 2 years; and bacteria without spore, for about 1 year. Even the *Meningococcus* which is difficult to preserve by ordinary method, can be maintained for 3 months in the incubator at 37℃. The advantage of the method is also easy to operate and no need of special equipment but the tube of culture must be kept erect.

3. Mixed with sand and soil

Removal of water can reduce the microbial metabolism. This method is applicable to the spores of molds, actinomycetes and bacteria and with a good effect. So it is widely used in the antibiotic industry. The procedures are to make the microorganisms be adsorbed to suitable carriers, such as soil or sand, and then dry the specimens. They can be preserved about 2 ~ 10 years. But the method is not applicable to the vegetative cells.

4. Ultra – cold frozen

Cultures can be frozen in low temperature refrigerator (– 20 ~ – 30℃, – 50 ~ – 80℃), dry – ice and alcohol (about – 70℃) and liquid nitrogen (– 196℃). For the method of freezing in liquid nitrogen, it is used for common microbes, also for what is difficult to preserve by freeze – drying, for example, mycoplasma, chlamydia, hydrogen bacteria, molds which is difficult to form spores, bacteriophages or some zooblasts. The advantage of the method is no variation but special equipments are required.

5. Freeze – drying

Freeze the microorganisms rapidly at low temperature (about – 70℃), then remove the water under reduced pressure with sublimation (vacuum desiccation). Make use of all factors that are benefit to culture preservation to keep the microorganisms in the condition of low temperature, dry and free from oxygen, so it is one of the most effective methods to preserve cultures.

The cultures that are preserved by freeze – drying can be maintained for several decades, but attention must be paid to their purity to prevent contamination for later use. Bacteria and yeasts are required to be cultivated to stationary phase, generally, 24 ~ 48 hours for bacteria, 3 days for yeasts, 7 ~ 10 days for actinomycetes and molds.

If cultures in logarithmic (log) phase are used, the survival rate will decrease. It is better to maintain spores instead of vegetative cells.

In the process of the latter two methods mentioned above, protecting agents are needed to prepare the cell suspension to keep the cells from damaging in adverse conditions like preventing the cells from the harm due to freezing or sublimation. The commonly used protecting agents are milk without grease, serum, saccharide, glycerol, dimethylsulfoxide and so on.

Some microorganisms such as virus, rickettsiaes and spirochetes etc, which cannot be cultivated in artificial media, can be preserved by host preservation. Specimens are subcultured into live animals, insects or chick embryos. This method is equivalent to subculturing.

Materials and Apparatus

1. Cultures

Bacteria, yeasts, actinomycetes and molds.

2. Media and other materials

Agar slant culture, degrease milk, sterile water, liquid paraffin, glycerol, sandy and soil, ice, salt, dry - ice, 95% alcohol, 10% HCl, anhydrous $CaCl_2$.

3. Apparatus

Sterile pipette, sterile dish, ampoules, sieve (40 mesh and 100 mesh), oilpaper, desiccator, vacuum pump, blowtorch, lower temperature refrigerator (-30℃), ultra - cold freezer, liquid nitrogen frozen incubator.

Procedures

1. Subculturing

(1) Slants at low temperature The cultures are inoculated to the agar slant and cultivated at optimal temperature. Keep in the refrigerator at 4℃ after it is well grown.

(2) Stab cultivation Transfer the semi - solid culture to a small tube (0.8cm 10cm) with 2 ~ 3cm away from the open of the tube. Using aseptic inoculating needle to transfer cells from agar slants and stab to 1/3 depth of the solid culture without touching the walls. After cultivating in optimal conditions, seal or plug the tube and then store it at 4 ~ 6℃.

2. Covered with liquid paraffin

(1) Place liquid paraffin in the autoclave for sterilization for 30min. (1.05kg/cm^2, 121℃)

(2) Inoculate the properly cultures to slant culture media and keep it well grown.

(3) Pour liquid paraffin into the agar slant, with 1cm higher than the peak of the

slant.

(4) Keep the tube erectly at 4℃.

When inoculated from liquid paraffin, note that sterilization of the inoculating loop can make the cultures flash together with the liquid paraffin.

3. Mixed with sand and soil

(1) Using 10% HCl to wash out organic impurity in river sand.

(2) Remove the acid solution, wash the sand to neutral, dry and sift out.

(3) Mill and sift out thin loess.

(4) Mix loess with sand in ratio of 1 to 3. Put them into small test tubes about 1cm high each and then sterilize.

(5) Carry on sterility test by taking one out of the ten tubes above randomly. Mix the content with broth culture media and place them in a 37℃ incubator for 48 hours. Resterilize if there are any bacteria.

(6) For inoculating, there are two methods as follows.

Dry method (for the preservation of actinomycetes and some of molds): transfer a loop of strong mature strains and inoculate to the sandy soil tube, thorough mixing. Do not inoculate the culture media into the tube.

Wet method: Select strong mature strains and wash them with 3 ~ 5ml of sterile water in order to get spore solutions. Add 10 drops (0.5ml) of these solutions to each sandy soil tube and mix them. Put these tubes into a vacuum desiccator to try to get rid of the water as fast as possible.

(7) Take one tube out of ten, from which inoculate some sands onto a slant, and then observe their growth.

(8) If there is no problem, seal the tubes and preserve them in the refrigerator or some dry places. Check their activity and purity once half a year.

(9) When it is claimed to use them, take some sands into the liquid culture.

4. Ultra – cold frozen in liquid nitrogen

(1) Prepare the ampoules Choose the ampoules made in silicon borate glass and preventing to crack due to sudden change of the temperature, holding 1.2ml. Tampon the ampoules and sterilize them.

(2) Prepare the protective agents and sterilize Prepare 10% glycerol or 10% dimethylsulfoxide as protective agents. Sterilize them as following: the former using autoclaving while the later using filtration. As for preservation of mold mycelium, it is necessary to add the protecting agents such as 10% glycerol in ampoules and sterilize for 20 min at 0.1MPa.

(3) Inoculation Prepare spore solution using 10% glycerol mixed with strains. Put them into sterile ampoules. As for mold mycelium, you may cut a piece of colony

block into the ampoule with protective agent. Seal the tubes and dip them into water to inspect whether they have holes.

(4) Freezing Freeze the ampoules slowly with 1℃ drop per minute till −30℃. (The crystal may be formed in cells to lower the livability of them if freeze them rapidly).

(5) Preservation Place the ampoules into liquid nitrogen incubator (gas nitrogen −150℃ and liquid nitrogen −196℃).

(6) Return to culture When use, take out the ampoules and put them in 38 ~ 40℃ water bath to ice – out rapidly till whole of them thawed. Open up the ampoules and transfer the content into the proper culture media.

5. Freeze – drying

(1) Prepare the ampoules Choose the ampoules with long neck and round bottom and made in hard glass. Immerse them in 2% HCl for 8 ~ 10 hours, then wash them with tap water for several times. The last, dry after wash them with distilled water for 2 ~ 3 times. Put the label with the name of strains and the date of inoculation into the ampoules. Tampon the ampoules and sterilize them at 0.1MPa for 30min.

(2) Prepare the protective agents – defatted milk Boil the fresh milk, defatted and then filter, after that, centrifugate it in 3000r/min for 15min. Repeat until the fat is removed completely. Sterilize it at 55.2kPa for 30min and then test the sterilization results.

(3) Prepare the strain Pay attention to the purity of the strain. For bacterial slant, add 2 ~ 3ml defatted milk into a slant to get concentrated solution. 0.2ml per ampoule (about 1/3 of the ampoule volume).

(4) Pre – freeze Place the ampoules in low temperature refrigerator or dry ice – alcohol (−25 ~ −40℃) to make the solution frozen.

(5) Vacuum drying Desiccate the ampoules while in the frozen state using a high vacuum to sublime the water, maintaining 6 ~ 8hours in 0.0267 ~ 0.0133 MPa.

(6) Seal the tubes Take out the ampoules, connect with special equipment to produce vacuum in them, then fuse the neck seal the ampoule in fire. Preserve them in refrigerator or in the dark at room temperature.

(7) Return to culture Firstly, sterilize the outer of the ampoules with 75% alcohol. Then, burn the upper of the ampoule and add drops of distilled water to make the tube crack. Inoculate the loose powder on the slant. Alternatively, pour the sterile liquid culture media into ampoule to dissolve the sample and then inoculate the bacteria solution to proper culture media.

This method is one of the most effective methods for preservation. It is used for common microbes, even for what is difficult to preserve, such as *Meningococcus* and

Gonococcus.

Results and Discussion

1. Which method can we use to preserve the bacteria simply?

2. Which method is used to preserve the bacteria without mutation for a long time?

3. Which method is used for spore – forming bacteria?

实验二十　微生物计数法

【实验目的】

1. 熟悉对单细胞微生物计数的原理及常见方法。
2. 了解血球计数板的构造，学习使用血球计数板进行微生物计数的方法。
3. 了解平板菌落计数法测定细菌浓度的基本原理和方法。

【实验原理】

单细胞微生物因其个体微小，其生长不是依据细胞大小，而是以在一定时间和条件下细胞数量的增加即群体的生长作为其生长的指标。

微生物群体生长情况可以通过测定单位时间里微生物细胞数目的增加或细胞物质的增加来评价。常见的测定方法有：①显微镜直接计数法；②测量细胞生物量，如细胞干重或湿重的测定；③间接计数法——平板菌落计数法；④细胞某种成分如蛋白质、核酸等的含量测定法；也有光电比浊法等。总之，测定方法很多，各有优缺点，工作中应根据具体情况加以选择。

本实验主要介绍测定细胞数目的显微镜直接计数法和平板菌落计数法。

1. 显微镜直接计数法

利用血球计数板在显微镜下直接计数，是一种常用的微生物计数方法。各种单细胞菌体的纯培养悬液如酵母菌、霉菌孢子和放线菌孢子等均可采用血球计数板计数。将经过适当稀释的菌悬液或孢子悬液放在血球计数板载玻片与盖玻片之间的计数室中，在显微镜下进行计数。由于计数室的容积一定（$0.1mm^3$），所以可以根据在显微镜下观察到的微生物数目来换算成单位体积内的微生物总数。此法的优点是直观、快速。缺点是不能区分死菌和活菌。由于此法计的是活菌体和死菌体的总和，故又称为总菌数计数法。

血球计数板是一块特制的厚载玻片，其上由四条槽构成三个平台。中间的平台又被一短横槽隔成两半，每一边的平台上各刻有一个方格网，中间的大方格即为计数室（图20－1）。

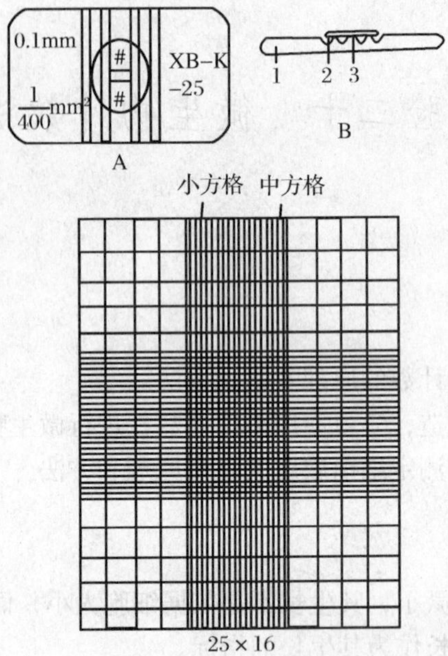

图 20－1　血球计数板构造

Fig. 20－1 The blood cell counter

A. 正面图；B. 纵切面图（1. 血球计数板；2. 盖玻片；3. 计数室）；C. 放大后的方格网（中央为计数室）
A. the right side；　B. the longitudinal section；　C. the grid Cin centrer

计数室的刻度一般有两种规格：一种是一个大方格分成 16 个中方格，而每个中方格又分成 25 个小方格；另一种是一个大方格分成 25 个中方格，每个中方格又分成 16 个小方格（图 20－2）。但两种规格的每一个大方格中的小方格总数都是 400（16×25）。每一大方格的面积为 $1mm^2$，盖上盖玻片后，载玻片与盖玻片之间的高度为 0.1mm，所以计数室的容积为 $0.1mm^3$。在计数时，如果是 25 个中方格，通常数 5 个中方格（即左上、左下、右上、右下和正中）共计 100 个小方格的总菌数，如果是 16 个中方格，通常数 4 个中方格（即左上、左下、右上、右下）共计 80 个小方格的总菌数，然后求得每个小方格中菌数的平均值，再乘上 400，就得出一个大方格中的总菌数，然后再换算成 1ml 菌液中的总菌数。

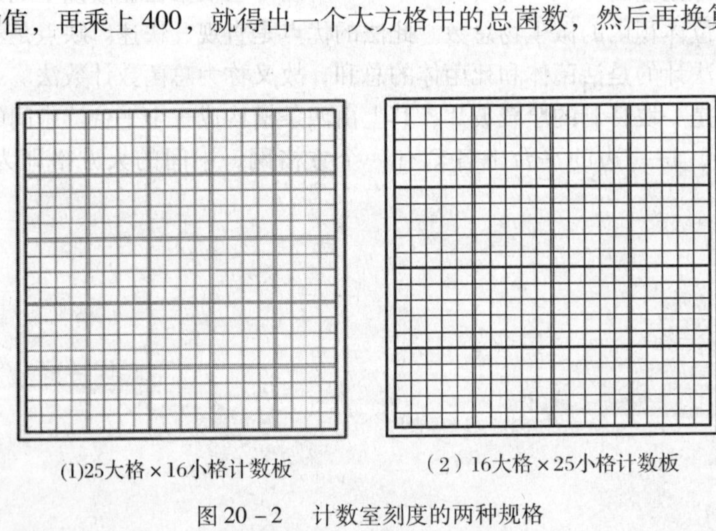

(1)25大格×16小格计数板　　（2）16大格×25小格计数板

图 20－2　计数室刻度的两种规格

Fig. 20－2　Two types of counter

可按照下列公式算出原孢子悬液的孢子浓度或原细胞悬液的细胞浓度。

样品中菌（或孢子）数（个/ml）=每小格的平均数×400×稀释倍数×10000 在这一公式中，10000 代表 1ml 的容积（即 $1000mm^3$）是一个计数室容积（$0.1mm^3$）的 10000 倍。

用于显微镜直接计数的计数器除血球计数板外，还有 Peteroff - Hauser 计菌器和 Hawksley 计菌器，它们都可用于对细菌、酵母菌、霉菌孢子等悬液的计数，原理基本相同。本实验介绍的血球计数板多适用于对较大的细胞计数，如酵母菌细胞、霉菌孢子以及动物细胞等，对个体较小的细菌不易观察和计数。后两种计数器计数室的容积是 $0.02mm^3$，即计数室面积是 $1mm^2$，而盖玻片与载玻片之间的距离是 0.02mm，因此可用油浸镜对细菌等较小的细胞进行观察计数。这两种计数器的使用方法可参看各厂家的说明书。

2. 平板菌落计数法

微生物在固体培养基上所形成的一个菌落是由一个单细胞繁殖而成的，称为一个菌落形成单位（colony forming unit，CFU），一个 CFU 即代表一个单细胞。平板菌落计数法计数时，先将待测样品作一系列稀释使其分散成单个细胞存在，再取一定量的稀释菌液接种到培养皿中（接种培养有两种方法：混合培养法和涂布培养法），使其均匀分布于平皿中的培养基内，经培养后，由单个细胞生长繁殖形成菌落，统计菌落数目，即可换算出样品中的含菌数。此法常用于某些成品检定（如杀虫菌剂），生物制品检定以及食品、水源的污染程度的检定以及药物微生物总量的检定等。这种计数法的优点是能测出样品中的活菌数。但平板菌落计数法的手续较繁，而且测定值常受各种因素的影响。

【实验材料】

1. 微生物及培养基

酵母菌（*S. cerevisiae*）悬液，大肠埃希菌悬液，肉汤培养基，营养琼脂培养基。

2. 仪器

血球计数板，显微镜等。

3. 其他

装有 4.5ml 蒸馏水的试管（灭菌的），无菌平皿，无菌吸管、无菌涂布棒等。

【实验方法】

1. 用血球计数板进行微生物计数

（1）准备待测菌　将酵母菌用无菌生理盐水制成浓度适当的菌悬液。

（2）镜检计数室　对计数室进行镜检，若有污物，则需清洗后才能进行计数。

（3）加样　将待测菌液由盖玻片与刻有计数室平台的接触处边缘滴一小滴（不宜过多），让菌液沿缝隙靠毛细渗透作用自动进入计数室（注意：取样时先要将菌液摇匀，加样时计数室不可有气泡产生）。静止 5 分钟。

（4）显微镜计数　将血球计数板置于显微镜载物台上，按照低倍镜、高倍镜使用方法，先用低倍镜找到计数室所在位置，然后换成高倍镜进行计数。凡是落在格左方和上方双线上的孢子或细胞都计算在内，而落在下方或右方双线上的孢子或细胞均不计算在内。

在计数前若发现菌液太浓或太稀，需重新调节稀释度后再计数。一般样品稀释度要求每小格内约有5~10个菌体为宜。

（5）清洗　使用完毕后，血球计数板先以75%酒精清洗消毒，再以蒸馏水淋洗并擦干。切勿用硬物洗刷，洗完后自行晾干或用吹风机吹干。

2. 平板菌落计数

（1）平皿、试管编号　取无菌平皿9套，分别用记号笔标明10^{-4}、10^{-5}、10^{-6}各三套。另取6支盛有4.5ml无菌水的试管，排列于试管架上，依次标明10^{-1}、10^{-2}、10^{-3}、10^{-4}、10^{-5}、10^{-6}。

（2）系列稀释　在实验室经一定时间培养的大肠埃希菌液体培养物作为待测样品，用无菌吸管吸取0.5 ml大肠埃希菌悬液加入10^{-1}的试管中，然后仍用此吸管将管内悬液吸吹3次，吸时伸入管底，吹时离开液面，使其混合均匀。另取一支吸管自10^{-1}试管吸取0.5 ml菌液加入10^{-2}试管中，吸吹3次（与上管操作相同），其余4管（10^{-3}、10^{-4}、10^{-5}、10^{-6}）的稀释过程依次类推。整个稀释过程如图20－3所示。

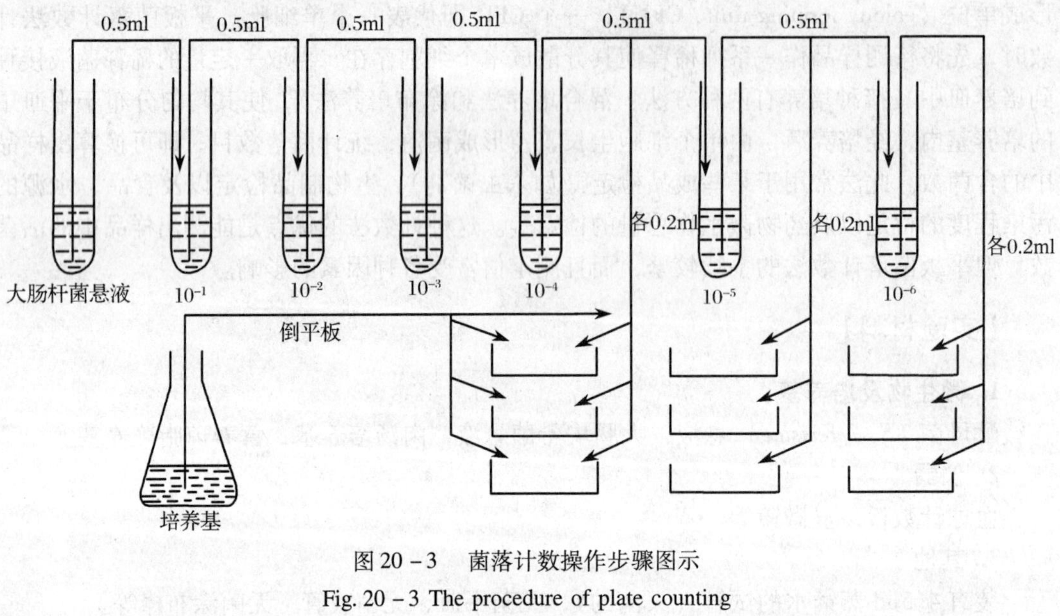

图20－3　菌落计数操作步骤图示

Fig. 20－3 The procedure of plate counting

（3）接种培养　有混合培养法和涂布培养法。

混合培养法：用3支无菌吸管分别精确地吸取10^{-4}、10^{-5}、10^{-6}的稀释菌液0.1ml（或0.5ml），分别加入对应编号的无菌培养皿中，在每个平皿中倒入熔化并恒温于45~50℃的营养琼脂培养基，迅速旋动平皿使菌液与培养基混匀（注意勿使液体溢出平皿外），静置待培养基凝固后，倒置于37℃培养。

涂布培养法：先将培养基熔化后倒入已编号的三套无菌平皿中，待凝固后，用无菌移液管分别吸取0.5ml稀释菌液对应加到上述已制备好的无菌平板表面，然后用无菌涂布棒将琼脂表面的菌液涂均匀（每个稀释度更换一个涂布棒）；将涂布过的平板置于桌面放置20分钟，使菌液渗入培养基中，最后将平板倒置于37℃培养箱中培养。

（4）计数　培养24小时后，取出培养基平板，记录同一稀释度的三个平板上的菌落数，取平均值后按下式计算出每毫升中总活菌数：

每毫升中总活菌数 = 同一稀释度平行三次的菌落平均数 × 稀释倍数 × 10（取 0.1ml 稀释菌液）（若取 0.5ml 稀释菌液，此处应为 ×2）

实验中若计数细菌、放线菌和酵母菌等菌落数，一般选择每个平板上长有 30 ~ 300 个菌落的稀释度较合适；若计数霉菌，选择每个平板上长有 30 ~ 100 个菌落的稀释度较合适。同一稀释度的三个平板中的菌落数不能相差很悬殊。由三个相邻的稀释度计算出的每毫升菌液中含菌数应相差不大，否则表示试验不精确。

【结果与讨论】

1. 将用血球计数板进行微生物计数的结果记录于表 20 – 1 中。

表 20 – 1　直接计数法结果（A 表示每小方格中的平均菌数；B 表示菌液稀释倍数）

	各中格中菌数					A	B	菌数/ml	平均值
	1	2	3	4	5				
第一室									
第二室									

2. 平板菌落计数的结果填入表 20 – 2。

表 20 – 2　平板菌落计数的结果

稀释度	10^{-4}				10^{-5}				10^{-6}			
菌落数	1	2	3	平均	1	2	3	平均	1	2	3	平均
活菌数/ml												

注：可以同时作两种接种培养法（混合培养法和涂布培养法），通过观察结果比较这两种方法各自的优缺点并分析实验误差。

3. 在混合培养法中，为什么熔化后的培养基要冷却至 45 ~ 50℃才能倒平板？
4. 试比较平板菌落计数法和显微计数法的优缺点。

20 Counting of Microorganisms

Purpose

1. To be familiar with the principles and methods of counting of microorganisms.
2. To understand the structure of blood cell counting plate and learn its procedures.
3. To be familiar with the principles and procedures of plate counting.

Principle

Unicellular microorganisms are tiny, therefore the index of their growth depends on the increase in cell numbers in given time and conditions instead of the sizes of them.

The growth of microorganisms can be assessed by the increase in cell numbers or that of cell substances. The common methods include: ① Direct counts of cell numbers under the microscope; ② Direct measurement of microbial biomass by weighing whole cells; ③ Indirect counts, such as using colony to estimate the cell numbers; ④ Indirect counts of microbial biomass relatively constant biochemical components of microbial cells, such as protein, nucleic acid and so on, also by measuring turbidity. Each of these methods has certain advantages and disadvantages, so the method selection varies in practice.

This experiment introduces the direct microscopic counting and plate counting.

1. Direct microscopic counting

The direct microscopic counts by using the blood cell counter is the commonly used method which is applicable to pure culture suspension of unicellular microorganisms, such as yeasts, the spores of molds and the actinomycetes. Directed microscopic counts is performed by spreading a measured volume of sample over a known area of a slide, counting representative microscopic fields, and reading the average number of cells. The advantage of the method is that it is direct and rapid to operate but it cannot distinguish between the dead and the viable microorganisms. The number of cells counted by this method is the summation of dead and viable cells, so it is also called total counting.

The blood cell counter is a specially – made thick slide which contains three platforms from four gabs. The middle platform is then divided into 2 parts by a short gab. There is a grid on each platform and the largest one in the middle is the counting chamber (Fig. 20 – 1) .

The slide is viewed in a microscope and the number of cells determined in an area are delimited by a grid. In the counting chamber shown, the largest grid with a total area of $1mm^2$ has 25 (or 16) large squares, each large square has 16 (or 25) small squares and the total number of small squares are 400 (16×25) and the height between the slide and the cover flake is 0.1mm (Fig. 20 – 2). With a glass flake, we can obtain a total volume of $0.1mm^3$. Counting the number of cells in the largest grid we can calculate the concentration of cells in the tested sample.

The concentration of the original sample can be calculated according to the following formulation.

The number of cells per ml = average number in each small square $\times 400 \times$ degree of dilution $\times 10000$.

Peteroff – Hauser counter and Hawksley counter are also used as the counting slides in direct microscopic counting besides blood ball counter. They are all used for counting the suspension of bacteria, spores of molds and actinomycetes. The blood ball counter is applicable for counting the larger cells such as yeast cells, mold spores and animal blood cells, it is difficult to observe and count the smaller ones. The volume of the two latter counting chambers is $0.02\ mm^3$ with the area $1mm^2$ and the distance between the slide and the cover slide 0.02mm. So immersion oil is suitable for smaller cells such as bacteria.

2. Plate counting

The most common procedure for the enumeration of bacteria is the viable plate counting. In this method serial dilutions of a sample containing viable microorganisms are plated onto a suitable growth medium. The suspension is either spread onto the surfaces of agar plates (spread plate method) or mixed with liquid agar, which is poured into plates and allowed to solidify (pour plate method). It is assumed that each bacterial colony arises from an individual cell that has undergone cell division, which is called colony forming unit (CFU), each CFU represents the progeny of a single cell. Therefore, by counting the numbers of colonies and considering for the original dilution factor, the number of bacteria in the original sample can be calculated. It is widely used in inspecting finished products (such as insecticide), biological product and the degree of contamination in food and water. The advantage of the method of viable plate counting is that it can detect the viable cells in sample. One of the major disadvantage is that the procedures are complicated and the results can be affected by many factors.

Materials and Apparatus

1. Microorganisms and cultures

Suspension of yeasts, *E. coli* suspension, nutrient broth, nutrient agar media and sterile water.

2. Apparatus

Counting chamber, microscope.

3. Others

Tube containing 4. 5ml sterile water, cover flake, sterile plate, sterile pipettes, sterile glass spreading rod, etc.

Procedures

1. The Microscopic Counting

(1) Prepare the sample to be determined Dilute and shake the yeast suspension to distribute cells evenly.

(2) Examine counting chamber Examine the counting chamber whether it is contaminated.

(3) Loading Transfer some of the yeast suspension to the contact edge of cover flake of counting slide; allow the solution to go into the counting chamber spontaneously by capillary osmosis. Be careful not to cause air bubble. Place it still for 5min.

(4) Microscopic counting Carefully place the counting chamber onto the microscope stage. Scan the ruled area of the counting chamber under low magnification and identify the large squares. Count the numbers of yeast cells in 5 (or 4) large squares. If cells fall on a line, include in your count those on the top and left hand lines and exclude those on the bottom and right hand lines. If the yeast cells are too dense to count, dilute your sample and start again. It is proper if there are 5 ~ 10 cells in each small square.

(5) Cleanout Cleanout the blood ball counting slide with 75% alcohol first and then rinse it with the distilled water. Do not brush it with hard matter. Dry it before conservation.

2. Plate Counting

(1) Label Plates and tubes Label nine plates of 10^{-4}, 10^{-5}, 10^{-6} three sets respectively. Label six tubes with 4. 5ml sterile water 10^{-6} through 10^{-1}.

(2) Serial dilution Dilutions are achieved by adding an aliquot of the sample of *E. coli* suspension to a sterile water dilution tube. If 0. 5ml of *E. coli* suspension is added to 4. 5 ml of sterile water, the dilution is 1 : 10; the other dilutions could also be achieved by adding sequentially dilution of the sample in series (Fig. 20 – 3).

(3) Inoculating

Pour plate method: Transfer 0. 1ml (or 0. 5ml) dilution from the tubes labeled 10^{-4}, 10^{-5}, 10^{-6} to plates with the same label respectively. The dilution in plate is mixed with a liquefied agar medium that has been cooled to 45 ~ 50℃. Allow the plates to stand until they are completely solidified, incubate the plates upside down at 37℃.

Spread plate method: Pour a liquefied agar medium to the three sterile plates labeled 10^{-4}, 10^{-5}, 10^{-6}. Transfer 0.5 ml dilutions with the same label to the above - mentioned agar plates respectively. The sterile glass spreading rod is then used to spread the suspension over the surface of the agar evenly. Place the plates on the table for 20 min to make the bacteria penetrate in the agar media. Incubate the plates upside down at 37℃.

(4) Counting After incubation, the numbers of colonies developed are counted and the concentration of microorganisms in the original suspension is calculated. For the plate - count data, calculate the concentration of bacteria in the original sample. For statistical reasons commonly the numbers of colonies between 30 and 300 are fit to use. The number of bacterial cells in the original sample is determined by multiplying the number of colonies on a dilution plate by the corresponding dilution factor. Generally replicates of each dilution are plated, and the mean count recorded.

The total number of living bacteria per ml = the average of bacteria colonies of the same degree of dilution × dilution factors × 10 (1ml diluted bacteria solution) (if 0.5 ml solution is used, multiplied by 2)

Results and Discussion

1. Record the result in the following table (Table 20 - 1).

Table 20 - 1 The results of microscopic counting

A: average number of bacteria each grid; B: degree of dilution.

Test No.	Number of bacteria each grid					A	B	Number/ml	Average
	1	2	3	4	5				
1									
2									

2. Fill out the following table (Table 20 - 2) with the counting results

Table 20 - 2 The results of plate counting

	10^{-4}				10^{-5}				10^{-6}		
	1	2	3	Average	1	2	3	Average	1	2	3
Number of colonies											
Viable cell No. /ml											

Note: Compare the two methods (spread plate method and pour plate method) through the results.

3. Why should we cool the melted agar to 45 ~ 50 ℃ and then pour it into plates?

4. Compare the advantages and disadvantages of the two methods: viable plate counting and the microscopic counting.

实验二十一　细菌的生长曲线测定

【实验目的】

1. 了解光电比浊计数法的原理，掌握光电比浊计数的操作方法。
2. 熟悉细菌生长曲线的特点，了解微生物在一定条件下生长繁殖的规律。

【实验原理】

在微生物学的研究和生产实践中，为了及时掌握发酵培养过程中的微生物生长情况，需要定时测定培养液中微生物的数量，以便对培养条件进行适当控制，获得最佳的培养物。比浊法是最常用的测定方法。比浊法的原理是在一定范围内，细菌悬液中的细胞浓度同光密度成正比，同透光度成反比。所以可用光电比浊计，在一定波长下来测定细菌悬液的光密度（即 optical density，OD），表示该菌在特定实验条件下的细菌相对数目，进而反映出其相对生长量。若用一系列已知菌数的菌悬液测定光密度，作光密度——菌数标准曲线，可以样品液的光密度从标准曲线中查出对应的菌数。本法优点是操作简便迅速，可连续测定并及时看到结果，但缺点是无法区分死菌和活菌。

若将一定量的微生物，接种在适合的新鲜液体培养基中，在适宜的温度、pH 值等条件下培养，以菌数的对数作纵坐标，生长时间作横坐标，做出的曲线叫生长曲线。一般生长曲线可分为延迟期、对数期、稳定期和衰亡期四个时期。

本实验采用比浊法测定大肠埃希菌的生长曲线，利用光电比色计测定菌悬液的光密度 OD 值，并将所测得的 OD 值与其对应的培养时间作图，即可绘出该菌在一定条件下的生长曲线。

如果我们将本实验中各培养时间的大肠埃希菌悬液同时采用稀释平板菌落计数法进行测数，可测定不同培养时间的活菌总数，以大肠埃希菌悬液光密度值为横坐标，以活菌数目为纵坐标，可绘制一条标准曲线。这样便可在测定任意培养时间菌悬液的光密度值后，在标准曲线上查出活菌总数。该方法较稀释平板菌落计数法节省时间，已经广泛应用在工业生产上。

【实验材料】

1. 微生物及培养基

大肠埃希菌培养液，牛肉膏蛋白胨培养基。

2. 仪器

72 型或 721 型分光光度计，恒温摇床。

3. 其他

锥形瓶，试管，无菌吸管，无菌生理盐水等。

【实验方法】

1. 接种及零时测定

试管法：取预先编号的 16 支分别装有 9.5ml 牛肉膏蛋白胨液体培养基的大试管，采用无菌操作技术用移液管向每个试管准确加入大肠埃希菌悬液 0.5ml，轻轻振荡混匀，迅速拿出 1 号管于分光光度计上测定 OD_{600}，此为零时的读数。另设空白对照组（不加大肠埃希菌悬液的液体培养基）用于调仪器零点。

摇瓶法：用无菌操作技术接种 5ml 已培养 6 小时的大肠埃希菌培养液于装有 95ml 肉汤培养基的锥形瓶中，震荡使菌混合均匀。立即取 2～3ml 于分光光度计上测定 OD_{600}，此为零时的读数。还需用相同的肉汤培养基作空白对照。

2. 培养

将接种后的其他 16 支大试管（或锥形瓶）置于摇床上，37℃恒温，200～220 转/分振荡培养。

3. 生长量测定

试管法：按下列数据记录表（表 21－1）中的时间间隔定时取出试管测定培养液的 OD_{600}。

摇瓶法：按表 21－1 的时间间隔从摇瓶中定时取 2～3ml 培养液测定 OD_{600}。

必要时可将培养液用空白培养基稀释后测定。（一般 OD 值应控制在 0.1～0.65 之间，精确度较高。）

4. 绘制生长曲线

以光密度（OD_{600}）为纵坐标，培养时间为横坐标，绘制大肠埃希菌的生长曲线。

【结果与讨论】

1. 数据记录

表 21－1　大肠埃希菌的生长曲线的测定数据

培养时间（h）	0	0.5	1	1.5	2	2.5	3	4
OD_{600}								
培养时间（h）	5	6	8	10	12	14	16	18
OD_{600}								

2. 光电比浊法的原理是什么？为什么说用比浊法测定的细菌生长只是表示细菌的相对生长状况？

3. 如果用平板活菌计数法制作生长曲线，你认为两个曲线有什么不同？

4. 在生长曲线中为什么会出现稳定期和衰亡期？

21 Bacterial Growth Curve

Purpose

1. To be familiar with the principle and procedures of turbidity measurement of microbial growth.

2. To understand the features of the bacterial growth curve and the rules of propagation under certain circumstance.

Principle

In the research of microbiology and the practice of manufacturing, in order to master the process of fermentation timely, it is necessary to determine the number of microorganisms, then properly control the conditions of cultivating and obtain the optimum cultures. Turbidimetry is the most commonly used method. The principle is that the concentration of cells in bacteria suspension is in proportion to the optical density, but in inverse proportion to the transmittance in given range. So we can use optical density (OD) of the bacteria suspension which is determined under a given wavelength to represent the relative number of bacteria in specific experiment conditions, and then reflect the relative growth. Determine the optical density by the bacteria suspension known number of bacteria, and then we can plot a standard curve of OD versus numbers of bacteria and count the corresponding number of the sample. The advantage of this method is that it is easy and rapid to operate but it cannot distinguish between dead and viable bacteria.

Bacterial population growth studies require inoculation of viable cells into a sterile broth medium and incubation of the culture under optimum temperature, pH, and gaseous conditions. Under these conditions, the cells will reproduce rapidly and the dynamics of the microbial growth can be charted by means of a population growth, which is constructed by plotting the increase in cell numbers versus time of incubation. The stages of a typical growth curve are Lag phase, logarithmic (log) phase, stationary phase, decline or death phase.

In this experiment, we introduce the turbidimetry to determine the growth curve of *E. coli*. We use a colorimeter to determine the OD of bacterial suspension, and plot a curve of OD versus time of cultivating, and then we get a growth curve under this condition.

If the numbers of *E. coli* cells at each incubation time are determined by diluted plate counting, total number of the viable bacteria versus the incubation time can be plotted to a standard curve. Thus, the number of viable bacteria at any incubation time can be obtained from the standard curve with the corresponding OD value. This method spends shorter time compared with the diluted plate counting method and as a result, it is widely used in industry.

Materials and Apparatus

1. Microorganisms and cultures

Broth culture of *Escherichia coli*, broth medium.

2. Apparatus

72 or 721 type spectrophotometer, shaker incubator.

3. Others

Conical flask, test tube, sterile pipette, sterile physiological saline.

Procedures

1. Inoculation and determination of zero

Tube method: Prepare 16 pre – labeled big test tubes with 9. 5 ml broth culture medium, add 0. 5 ml of *E. coli* suspension to each tube respectively, and mix them gently. Take out the 1 test tube immediately to determine the initial OD_{600} (t_0). Select a tube the sterile broth medium without any *E. coli* as a blank.

Shake flask method: Inoculate 5ml broth culture of *E. coli* to 95ml nutrient broth in conical flask; shake it to mix the culture, the initial OD_{600} (t_0) is determined, using the sterile broth medium as a blank.

2. Cultivating

Place the other 16 inoculated big test tubes (or flask) in the shaking incubator and then cultivate at 200 ~ 220r/min at 37℃.

3. Determine the growth

Tube method: Take out the tubes to measure OD_{600} according to the following table (Table 21 – 1), the culture solution can be diluted with the medium when necessary (it is precise when the range of OD is controlled in the range of 0. 1 ~ 0. 65).

Shake flask method: According to the following table (Table 21 – 1), aseptically transfer 2 ~ 3ml aliquot of the culture to a cuvette and determine its optical density.

4. Plot the growth curve

Plot the growth curve of *E. coli* (OD as ordinate and time of cultivating as abscissa).

Results and Discussion

1. Records the following table:

Table 21 – 1 The results of growth curve of *E. coli*

Time of cultivating (h)	0	0.5	1	1.5	2	2.5	3	4
OD_{600nm}								
Time of cultivating (h)	5	6	8	10	12	14	16	18
OD_{600nm}								

2. What is the principle of turbidimetry? why the growth of the bacteria estimated by this method only suggests the relative growth status?

3. Compare the curve plotted by turbidimetry and that of viable plate counting; what is the difference between them?

4. Why do the two stages as stationary phase and decline phase in growth curve occur?

第四部分　细菌的生理生化反应

实验二十二　淀粉水解和明胶液化试验

异养型微生物依赖于周围环境的有机物生长，而许多有机物由于分子量太大、结构过于复杂，不能被微生物吸收利用。有些微生物可以分泌水解性酶类到细胞外，在体外将大分子有机物分解为它们的结构单元或亚基，再加以吸收利用。不同类型的异养微生物可以水解的有机物是不同的，借此可以对微生物进行分类鉴定。

【实验目的】

了解不同的微生物对淀粉（多糖）和明胶（蛋白质）的分解能力，说明微生物具有能分泌不同酶的特性。

【实验原理】

淀粉是葡萄糖的多聚物，微生物对淀粉的利用依赖于分泌到细胞外的淀粉酶，淀粉首先被分解成糊精，再进一步分解成麦芽糖和葡萄糖：

淀粉→糊精→麦芽糖→葡萄糖

淀粉和碘可进行反应生成蓝色的化合物，而淀粉的水解产物：糊精、麦芽糖和葡萄糖却不会生成蓝色化合物，据此可以用碘试剂来判断细菌是否具有分解淀粉的能力。

蛋白质是氨基酸的多聚物，蛋白质如明胶的水解过程为：

明胶→朊蛋白→蛋白胨→多肽→氨基酸

明胶酶是有些微生物分泌的一种蛋白酶，催化蛋白质分解。明胶液化试验用于判断微生物是否具有分解蛋白质的能力。将待检微生物穿刺接种于明胶培养基，经过培养，如细菌分解明胶就会发生明胶液化现象。

【实验材料】

1. 细菌

枯草杆菌、金黄色葡萄球菌和大肠埃希菌斜面培养物。

2. 试剂

淀粉琼脂平板、碘试剂、明胶高层培养基。

3. 其他

接种环。

【实验方法】

1. 淀粉水解试验

（1）将淀粉琼脂平板分成三个区，并在底部做好标记。

（2）使用无菌操作，在各区划线接种适宜的培养物，倒置培养。

（3）置37℃恒温箱中培养48小时后，在平皿上滴加碘液，无菌处平皿呈蓝黑色。若细菌生长线周围变成透亮的无色，则为淀粉试验阳性（+），说明该菌能够产生淀粉酶，将淀粉水解；若细菌生长线四周仍为蓝黑色，表明淀粉不被水解，淀粉水解试验为阴性（-）。

2. 明胶液化试验

（1）取明胶高层培养基三支，通过穿刺接种法，一支接种大肠埃希菌，另一支接种枯草杆菌，余下一支为对照。

（2）置37℃恒温箱中培养24~48小时。

（3）将试管置于冰箱或冰浴中30~60分钟。

（4）观察结果，明胶培养基呈液化状态者为阳性（+）；无液化现象发生者则为阴性（-）。

【结果与讨论】

1. 淀粉水解试验结果

细菌	枯草杆菌	金黄色葡萄球菌
淀粉水解		
对照		

2. 明胶液化试验结果

细菌	枯草杆菌	大肠埃希菌
明胶液化		
对照		

3. 什么是水解？它在本试验中的作用是什么？

4. 碳水化合物、脂肪以及蛋白质各自的结构单元是什么？

PART FOUR Biochemical Tests of Bacteria

22 Hydrolysis of Starch and Protein

Heterotrophic microorganisms are dependent on organic molecules in their environment. Many types of organic molecules are too large and complex to be absorbed by the cells. The cells can use these molecules when their extracellular hydrolytic enzymes are secreted into the environment, digesting the molecules into their component subunits, and transporting these subunits across the membrane into the cells. Not all types of heterotrophic microorganisms are equally adept at hydrolyzing specified organic compounds, and the tests you will perform in this exercise are often valuable in identifying microbial species or strains.

Purpose

1. To demonstrate the extracellular hydrolysis of starch and gelatin.
2. To understand that the microbes have the ability to secrete enzymes to hydrolyze the macromolecules around the cells.

Principle

The starch is a polymer of glucose; starch hydrolysis by bacteria depends on the extracellular enzymes called amylases. The starch molecule is first hydrolyzed to smaller polysaccharide molecules, it is so called dextrin. The dextrin is then hydrolyzed to maltose, and the maltose is hydrolyzed to glucose, the end product of the digestive process:

$$\text{Starch} \rightarrow \text{dextrin} \rightarrow \text{maltose} \rightarrow \text{glucose}$$

Iodine interacts with starch and produces a dark blue complex. The hydrolytic products of starch – dextrin, maltose and glucose – do not interact with iodine in this manner. Therefore we can use an iodine solution to determine whether a given bacterial strain is capable of hydrolyzing starch.

Protein molecules are composed of amino acid subunits. Hydrolysis of a large protein molecule such as gelatin, proceeds through smaller and smaller amino acid chains and finally to amino acids themselves:

gelatin →proteoses →peptones →peptides →amino acids

Gelatinase, which is produced by many microorganisms, catalyzes the hydrolysis of the gelatin; the measurement of gelatin hydrolysis is a measure of the activity of this enzyme. The hydrolysis of gelatin produces soluble carbohydrates that are readily metabolized as a source of carbon and energy.

Materials and Apparatus

1. Culture

Broth cultures of *Escherichia coli*, *Bacillus subtilis* and *Staphylococcus aureus*

2. Reagents

Starch agar plate, Iodine solution, sterile nutrient broth tubes supplemented with 12% gelatin.

3. Apparatus

Inoculating needle.

Procedures

1. Starch Hydrolysis

(1) Mark the bottom of the starch agar plate into three sectors.

(2) Using aseptic technique, inoculate each sector with the appropriate culture, making a wavy line. Invert the plate and place it in the designated area.

(3) After 48h, flood the surface of the plate with iodine solution. Note that the surface of the agar is not occupied by bacterial growth become to dark blue. If a clear zone appears around a line of bacterial growth, hydrolysis has occurred. If the agar immediately adjacent to a line of bacterial growth is dark blue, hydrolysis has not occurred.

(4) Record the results and discussions in the table.

2. Gelatin Hydrolysis

(1) Inoculate one nutrient broth - gelatin tube with each microorganism. Leave one tube uninoculated as a control. Label the tubes.

(2) Incubate at 37℃ for 24 ~48h.

(3) place the tubes in a refrigerator or an ice bath for 30 ~60min.

(4) Check each of the tubes for solidification. If solid, the reaction is negative, indicating the lack of gelatin hydrolysis.

(5) Record your results and discussions.

Results and Discussion

1. Results of Starch Hydrolysis

Bacteria	*Bacillus subtilis*	*Staphylococcus aureus*
starch		
control		

2. Results of Gelatin Hydrolysis

Bacteria	*Bacillus subtilis*	*Escherichia coli*
gelatin		
control		

3. What is hydrolysis? What role does hydrolysis play in the tests you performed in this exercise?

4. Distinguish among carbohydrates, fats and proteins. What are their subunits called, respectively?

实验二十三 糖发酵试验

【实验目的】

1. 测试微生物利用单糖（如葡萄糖）和双糖（如乳糖、蔗糖）的能力。
2. 理解不同的微生物具有利用不同碳源的酶系的能力。

【实验原理】

微生物具有不同的利用各种碳源的能力，其原理在于不同微生物具有不同的酶系。微生物利用碳源能力和结果的不同可用于鉴定微生物。

溴甲酚紫是一种酸碱指示剂，在 pH 中性时为紫色，碱性时为深红色，而在酸性时呈现黄色。试验时，在各试验管中加一倒置小管，称为杜氏管（Durham tube），分装入试验用培养基，高压灭菌，培养基被压进杜氏管，并赶走管内气体。微生物在生长过程中产生的气体滞留在管内，可以观察到。当溴甲酚紫的颜色由紫色变为黄色时，表明微生物利用碳源产生了酸性物质。

微生物在进行碳源代谢时可以产生不同的代谢产物。有些产物为酸性物质——如醋酸、甲酸以及乳酸等。酸性物质的积累有时会超出培养基的缓冲范围，导致 pH 下降，溴甲酚紫的颜色由紫色转为黄色。在产酸过程中，有时伴随气体的产生，这可以在杜氏管中的气泡来验证；若碳源代谢的终产物为中性化合物，既无颜色变化也无气体产生，表明此时的代谢较为复杂。

【实验材料】

1. 细菌

大肠埃希菌和伤寒杆菌培养物。

2. 试剂

蛋白胨水培养基，葡萄糖，乳糖，麦芽糖，甘露醇，蔗糖，溴甲酚紫指示剂。

3. 其他

接种环、试管、杜氏小管。

【实验方法】

1. 在试验管上标记好试验用菌的名称（各 6 个试管）。
2. 使用无菌操作，将试验用菌接入各个试管，第六管为对照，不接种。
3. 将试验管置于培养箱中，直立放置培养。
4. 48 小时后和 5 天后检查培养状况。
5. 记录试验结果：NR 表示无反应或结果复杂；A 表示产酸不产气；“AG”表示产酸

产气；“B”表示产生碱性物质。

【结果与讨论】

1. 将实验结果填入下表。

生化反应	细菌	现象与结果					原理
		葡萄糖	乳糖	麦芽糖	甘露糖	蔗糖	
糖发酵试验	大肠埃希菌						
	伤寒杆菌						

2. 本试验中对照管有什么意义？

23 Utilization of Carbohydrates

Purpose

1. To test the ability of bacterial species to use the monosaccharide sugar glucose and the disaccharide sugars lactose and sucrose.

2. To understand that different microorganism has different enzymes to use carbohydrates.

Principle

Organisms differ in their ability to use different carbohydrates. These differences are useful in the characterization of a microorganism. The phenomena lies on these microorganisms have different enzymes to use carbohydrates.

Bromcresol purple is a pH indicator that is purple when neutral, deep red when basic, and yellow when acidic. Each tube contains an inverted vial, or Durham tube. Such a broth tube is prepared by dispensing the medium into the tube. The pressure of autoclaving forces the medium into the vial, which displaces the trapped air. Subsequent gas production in the medium by a bacterial strain can be detected, because some of the gas becomes trapped in the vial and displaces the medium. Acid production is indicated by a change in color from red to yellow.

Organisms may produce a variety of end products in carbohydrate metabolism. Some of these products are acids – for example, acetic, formic, and lactic acids. If acids accumulate in sufficient quantity to overcome the buffers present in the medium, the pH is lowered. A change in color from purple to yellow indicates this alteration in pH. Acid production may or may not be accompanied by gas production. If acid and gas are produced, the medium turns yellow and gas accumulates in the inverted vial. In some cases the end products of carbohydrate metabolism are neutral. In this event, no color change occurs nor does gas accumulate in the vial. The only indication of growth is turbidity.

Materials and Apparatus

1. Bacteria

Broth cultures of *Escherichia coli* and *Bacterium typhi.*

2. regent

Carbohydrates broth, glucose, lactose, maltose, mannitol, sucrose, bromcresol purple.

3. Apparatus

Inoculating loop, ten tubes of bromcresol purple carbohydrates broth containing Durham tubes.

Procedures

1. Mark each set of 5 tubes with the names of your 2 cultures (5 tubes of each culture)

2. Using aseptic technique, inoculate each of the 5 tubes. as control, the sixth tube in each case is not to be inoculated.

3. Place the tubes in the assigned location, make sure they are in an upright position.

4. Examine the tubes after 48 hours and again after 5 days.

5. Record the results and discussions here and, in the case of unkown, on the data sheet. Use the following key:

NR (for "no reaction"): tube shows turbidity only.

A (for "acidic"): yellow tube, no gas production.

AG (for "acid and gas"): yellow tube, gas production.

B (for "basic"): darker red color than control.

Results and Discussions

1. Describe the results of utilization of carbohydrates

Carbohydrates	glucose	lactose	maltose	mannitol	sucrose
E. coli					
B. typhi					

2. Why is it important to maintain control tubes for the tests performed in this exercise?

实验二十四　IMViC 试验

【实验目的】

1. 掌握微生物鉴定中常用的几种生化反应原理及结果判断方法。
2. 熟悉微生物生化反应中各种培养基的设计和用途。

【实验原理】

IMViC 是以下四个试验的缩写：吲哚试验（indole production，I）、甲基红试验（methyl red test，M）、V－P 试验（Voges－Proskauer test，V）和枸橼酸盐利用试验（citrate utilization test，C），字母“i”是为了发音的需要加入的。

IMViC 试验常用于革兰阴性的肠道细菌检测中。如产气杆菌和大肠埃希菌在许多测试中反应很相似，极其容易混淆。IMViC 则可以区分产气杆菌属和大肠埃希菌属的微生物。

1. 吲哚试验

色氨酸几乎存在于所有蛋白质中，有些细菌可以将色氨酸分解为吲哚，吲哚在培养基中的积累可以由寇氏试剂（Kovacs' reagent）检测出来。试验操作必须在 48 小时内完成，否则吲哚进一步代谢，会导致假阴性的结果。

寇氏试剂包含三种成分——盐酸、异戊醇和对二甲基氨基苯甲醛，每种试剂均有其作用；醇用于浓缩分散在培养基中的吲哚；对二甲基氨基苯甲醛可以和吲哚反应形成红色的化合物，该反应必须在酸性条件下完成，盐酸的作用就是制造酸性环境。一旦指示剂的颜色变为红色，就表明吲哚试验为阳性（+）。

2. 甲基红试验

大肠埃希菌和产气杆菌利用葡萄糖时有所不同，大肠埃希菌接种到诸如 MRVP 培养基中时，将产生一些酸性物质，导致 pH 下降。而产气杆菌利用葡萄糖时则产生中性物质，培养基的 pH 没有显著变化。

3. V－P 试验

产气杆菌利用葡萄糖时产生的中性物质之一就是乙酰甲基甲醇，大肠埃希菌并不产生此物质。V－P 试验可用于特异性地检测乙酰甲基甲醇。V－P 试验和甲基红试验一起，是检测大肠埃希菌和产气杆菌的最有效的方法。

4. 枸橼酸盐利用试验

另一个可以区分大肠埃希菌和产气杆菌的培养基是枸橼酸盐琼脂。若以枸橼酸盐作为唯一的碳源制备培养基，大肠埃希菌不能在上面生长，而产气杆菌却可以生长得特别好。而且产气杆菌代谢产生的终产物为碱性，最终导致培养基 pH 的显著上升。指示剂溴百里酚蓝（brom thymol blue）可以检测到这一变化，pH 中性时溴百里酚蓝为绿色，当 pH 达到 7.6 时，颜色转为深蓝。除此之外，枸橼酸盐利用试验也可以用于某些肠道致病菌的检查。大多数的沙门菌可以利用枸橼酸盐，但是伤寒沙门菌和所有志贺菌却不利用。

【实验材料】

1. 细菌

大肠埃希菌培养物；产气杆菌培养物。

2. 试剂

寇氏试剂、蛋白胨－水－磷酸盐培养基；甲基红试剂；6% α－萘酚；40% KOH 溶液；西蒙枸橼酸盐琼脂斜面。

3. 其他

接种环等。

【实验方法】

1. 吲哚试验

（1）在2支装有试验用培养基的试管上标记好试验菌名称，另一管作为对照。

（2）按照无菌操作，接种于试验菌培养基中。

（3）37℃培养48小时后，取出试验管，每管加入10滴寇氏试剂，在手掌中搓动试管，使管内液体混合均匀，置于试管架上5分钟后，观察，寇氏试剂由黄色转为红色表示有吲哚存在，为试验阳性（+）。

（4）记录试验结果。

2. 甲基红试验

（1）在2支装有试验用培养基的试管上标记好试验菌名称，另一管作为对照。

（2）按照无菌操作，接种于试验菌培养基中。

（3）37℃培养48小时后，取出试验管。取两个空试管，每管加入5ml培养基（此时无须无菌操作）。再加入5滴甲基红试剂，若甲基红试剂由黄色转为红色表示培养基为酸性，为试验阳性（+）。

（4）记录试验结果。

3. V－P 试验

（1）在2支装有试验用培养基的试管上标记好试验菌名称，另一管作为对照。

（2）按照无菌操作，接种于试验菌培养基中。

（3）37℃培养48小时后，取出试验管。加入0.5ml α－萘酚溶液以及0.5ml 40% KOH，静置5分钟，若管内颜色转为红色为V－P试验阳性（+），如果所有试管均无红色产生，应稍微加热后，再看试验结果。

（4）记录试验结果。

4. 枸橼酸盐利用试验

（1）在3支装有试验用西蒙氏培养基的试管上标记好试验菌名称，另一管作为对照；

（2）按照无菌操作，接种于试验菌培养基中。

（3）37℃培养48小时后，取出试验管，观察，若管内颜色转为深蓝色为枸橼酸盐利用试验阳性（+），若试管仍为绿色，为试验阴性（－）。

（4）记录试验结果。

【结果与讨论】

1. IMViC 试验结果

IMViC	细菌	试验现象	结果	原理
吲哚试验	大肠埃希菌			
	产气杆菌			
甲基红试验	大肠埃希菌			
	产气杆菌			
V. P 试验	大肠埃希菌			
	产气杆菌			
枸缘酸盐利用试验	大肠埃希菌			
	产气杆菌			

+为试验阳性；-为试验阴性。

2. 说明寇氏试剂的主要成分以及各自的作用。

3. 说明 MRVP 培养基的主要成分，各自的功能是什么？字母 MRVP 代表什么？

4. 在 V-P 试验中利用何种细菌代谢的中间物进行检测？它来源于何种底物的代谢产物？

5. 在西蒙枸橼酸盐斜面加入 5g/L 的葡萄糖，会改变该培养基的作用吗？

24 IMViC Series

Purpose

1. To understand the principle of biochemical tests and the judgment of these tests
2. To be acquainted with the design and purpose of test mediums

Principle

The acronym IMViC refers to a series of four tests: indole production (I), the methyl red test (M), the Voges – Proskauer test for the production of acetyl – methylcarbinol (V), and the citrate utilization test. (The "i" in IMViC is included merely for ease of pronunciation). The IMViC series is valuable for the information it can provide, but it is especially useful in differentiating gram – negative enteric bacteria. For example, *Enterobacter aerogenes* is often confused with *Escherichia coli* because of their similar reactions to various tests. This confusion sometimes arises in analyzing the bacterial content of water. The IMViC series may be run to make a distinction between members of the genus *Escherichia* and members of the genus *Enterobacter.*

1. Indole production

The amino acid tryptophan is almost universally present in protein. Some bacteria can degrade tryptophan to indole. Indole accumulates in the medium, where it can be detected with Kovacs' reagent. It is important to perform the test within 48h, because after this time indole begins to be metabolized and the test often yields false negative results and discussions.

Kovacs' reagent contains three ingredients: hydrochloric acid, isoamyl alcohol, and para – dimethyl – amina – benzaldehyde. Each ingredient plays a critical role in the test for indole. The alcohol concentrates indole, which is dispersed throughout the medium. Para – dimethyl – amina – benzaldehyde reacts with indole to form a red complex. Finally, this reaction occurs only at a low pH, which explains the function of hydrochloric acid. Thus a positive indole test involving Kovacs' reagent is indicated by a color change from yellow to red.

2. The Methyl Red Test

Escherichia and *Enterobacter* differ in their utilization of glucose. If *Eschericia* is inoculated into a medium containing glucose, such as MRVP broth, several acids are produced. As a result, there is a marked decline in pH. By contrast, *Enterobacter* does not produce acids in its metabolism of glucose. Instead, the end products are neutral and there is no significant change in pH.

3. The Voges – Proskauer Test

One of the neutral substances produced by *Enterobacter* in its metabolism of glucose is acetylmethylcarbinol. *Escherichia* does not produce this substance. The Voges – Proskauer test is specific for acetylmethylcarbinol. Together with the methyl red test, it is most useful in distinguishing *E. coli* from *E. aerogenes.*

4. Citrate Utilization

An additional medium that can be used to distinguish *Escherichia* and *Enterobocter* is citrate agar. If citrate (a salt of citric acid) is the sole carbon source in a medium, *Eschichia* can not grow. By contrast, *Enterobacter* grows quite well on citrate. Furthermore, the end products of *Enterobacter's* citrate metabolism are basic and raise the pH of the medium significantly. Incorporation of the pH indicator brom thymol blue provides a ready means of measuring the rise in pH. (Brom thymol blue is green at a neutral pH but changes to a deep blue at pH 7.6). In addition to its value in distinguishing *Escherichia* and *Enterobacter*, the citrate utilization test may be used to differentiate among certain intestinal pathogens. Most members of the genus *Salmonella* utilize citrate, but *S. typhi* and all members of the genus *Shigella* do not.

Materials and apparatus

1. Culture

Broth cultures of *Escherichia coli* and *Enterobacter aerogenes.*

2. Reagent

Bottle of Kovacs' reagent, tubes of tryptone broth, tubes of MR – VP broth, methyl red indicator, Bottle of alpha naphthol, Bottle of potassium hydroxide (KOH) solution, Simmons' citrate agar slants.

3. Apparatus

Inoculating loop.

Procedures

1. Indole production

(1) Mark 2 of the tryptone broth tubes with the names of two cultures.

(2) Using aseptic technique, inoculate the tubes and place them in the appropriate

location for incubation. The third tube, which serves as a control, is also to be incubated also.

(3) Remove the tubes from incubation after 48h, and add 10 drops of Kovacs' reagent to each tube. Rotate the tubes between your palms to mix the contents. Place them in a test tube rack and let them stand for 5min. (Kovacs' reagent rises to the top.) A change in the color of Kovacs' reagent from yellow to red indicates the presence of indole.

(4) Record your results and discussions.

2. The Methyl Red Test

(1) Mark 2 of the MRVP broth tubes with the names of two cultures.

(2) Using aseptic techniques inoculate the tubes and place them in the appropriate location for incubation. The third tube, which serves as a control, is also to be incubated also.

(3) Remove the tubes from incubation after 48h. Mark the 2 empty tubes, and pour 5ml of the medium from each culture into an empty tube. Because the medium is "spent," you need not make this transfer aseptically. However, take care not to spill any of the culture medium. Do not discard the remainder of the spent medium; you will use it in the Voges – Proskauer test. Add 5 drops of methyl red indicator to each of the 2 tubes that you just poured. If the pH of the medium is acids, the color of the reagent will remain red. If the pH of the medium is neutral or basic, the color will change to yellow.

(4) Record your results and discussions.

3. The Voges – Proskauer Test

(1) Mark 2 of the MRVP broth tubes with the names of two cultures.

(2) Using aseptic techniques inoculate the tubes and place them in the appropriate location for incubation. The third tube, which serves as a control, is also to be incubated also.

(3) Remove the tubes from incubation after 48h. Add 0.5ml of alpha naphthol and 0.5ml of potassium hydroxide solution to each of the original culture tubes (now containing 5ml of spent medium). Allow the tubes to stand for 5min. A red color indicates the production of acetylmethylcarbinol. Note: If none of the tubes shows a red color after 5min, subject the *E. aerogenes* tube to mild heating, which enhances formation of color. Subject the other tubes to this same degree of heating.

(4) Record your results and discussions.

4. Citrate Utilization

(1) Mark 3 of the Simmons' citrate agar tubes with the names of 3 cultures.

(2) Using aseptic technique, inoculate the tubes by the stab technique and incubate

them at 37℃. The third tube serves as a control; stab the agar with a sterile needle.

(3) After 48 hours, remove the tubes from the incubator and observe the color. A blue color indicates the ability to utilize citrate as a carbon source. A green color indicates the inability to utilize citrate as a carbon source.

(4) Record your results and discussions.

Results and Discussions

1. Record the results of IMViC on the data sheet.

IMViC	Indole	Methyl Red	Voges – Proskauer	Citrate Utilization
E. coli				
E. aerogenes				

2. Name the key ingredients in Kovacs' reagent. What is the special role of each ingredient?

3. Name two key ingredients in MRVP broth. What is the function of each these ingredients? What do the letters MRVP represent?

4. What compound is the Voges – Proskauer test designed to detect in a growth medium? What substrate is metabolized to produce this compound?

5. Suppose you add 5 grams of glucose per liter to Simmons' citrate agar. Would this change make any difference in the functioning of this medium?

实验二十五　硫化氢产生试验和触酶试验

【实验目的】

1. 掌握微生物鉴定中常用的几种酶催化的反应原理及结果判断方法。
2. 熟悉微生物生化反应中各种培养基的设计和用途。

【实验原理】

硫化氢（Hydrogen sulfide，H_2S）：某些微生物在分解半胱氨酸等含硫氨基酸时脱硫产生的。硫化氢的产生是半胱氨酸转化为丙酮酸和氨这一系列反应的第一步，可以由醋酸铅检测出来，硫化氢遇到醋酸铅，可以形成黑色的沉淀。

触酶催化过氧化氢分解为水和氧，大多数需氧微生物可进行该过程。触酶的功能是将代谢过程产生的毒性的过氧化氢除去，以保护细胞不被伤害。当过氧化氢加入到微生物或菌体上，将有氧气气泡产生，以此证明触酶的存在。

【实验材料】

1. 细菌

甲型副伤寒杆菌培养物；乙型副伤寒杆菌培养物；枯草杆菌培养物；梭杆菌培养物。

2. 试剂

醋酸铅营养琼脂高层；过氧化氢溶液（3% ~10%）。

3. 其他

接种针；灭菌牙签；灭菌枪头；载玻片；体视镜等。

【实验方法】

1. 硫化氢产生试验

（1）使用接种针。按照无菌操作，将细菌接入醋酸铅琼脂高层。一支穿刺接种甲型副伤寒杆菌，另一支穿刺接种乙型副伤寒杆菌，注明菌名。

（2）置37℃恒温箱中培养24 ~48 小时。

（3）培养结束后，观察结果时，有黑褐色硫化铅者为阳性（+），无此现象者为阴性（-）。

（4）记录实验结果。

2. 触酶试验

（1）使用接种环或无菌牙签将枯草杆菌和梭杆菌平板或斜面上的菌体取下。

（2）将菌体涂于载玻片上。

（3）用移液枪将一滴过氧化氢溶液加于菌体之上。

（4）观察有无气泡产生，若有，则为触酶试验阳性。有时产生气泡较小，观察较困难，可使用体视镜辅助观察。

（5）记录结果。

【结果与讨论】

1. 硫化氢产生试验结果

生化反应	细菌	实验现象	结果	原理
硫化氢产生试验	甲型副伤寒杆菌			
	乙型副伤寒杆菌			

+为试验阳性；－为试验阴性。

2. 触酶试验结果

生化反应	细菌	实验现象	结果	原理
触酶试验	枯草杆菌			
	棱杆菌			

+为试验阳性；－为试验阴性。

25 Hydrogen Sulfide Production and Catalase Test

Purpose

1. To understand the principle of enzymes tests and the judgment of these tests.
2. To be acquainted with the design and purpose of test media.

Principle

Hydrogen sulfide (H_2S) is produced from cysteine by certain bacteria that synthesize the enzyme cysteine desulfurase. The production of H_2S takes place in the first of a series of reactions in which cysteine is converted to pyruvic acid and ammonia. The production of hydrogen sulfide can be detected by including lead acetate in the medium; when H_2S reacts with lead, a black precipitate (PBS) forms.

Catalase is an enzyme that catalyzes the decomposition of hydrogen peroxide (H_2O_2) to water and gaseous oxygen. Most aerobic microorganisms possess catalase. The function of catalase is to remove toxic hydrogen peroxide that forms during the oxidation – reduction reactions that are coupled with oxygen in respiratory metabolism. The presence of catalase is important for aerobic growth because it prevents the accumulation of toxic metabolites that would otherwise kill the cell. The presence of catalase is demonstrated when hydrogen peroxide is added to a colony or loopful of bacteria and bubbles of oxygen are released from the surface.

Materials and Apparatus

1. Culture

Broth cultures of *Escherichia coli*, *Proteus vulgaris*, *Bacillus subtilis* and *Clostridium sp.*

2. Reagent

Butt tubes of lead acetate – nutrient agar; Hydrogen peroxide (3% ~ 10% solution).

3. Apparatus

Inoculating loop; Sterile wooden applicator sticks or toothpicks; Pasteur pipettes; Microscope slides; Binocular (dissecting) microscope.

Procedures

1. Hydrogen sulfide production

(1) With the inoculating needle, aseptically transfer each culture to a tube of lead acetate nutrient agar medium. Stab the agar butt with the inoculating needle to inoculate the subsurface regions of the agar.

(2) Incubate the tube with *Salmonella paratyphi A* and *Salmonella paratyphi B* at 37℃ for 24 ~ 48h.

(3) observe the tubes after incubation. The presence of a black precipitate along the line of inoculation indicates the production of hydrogen sulfide.

(4) Record your results and discussions.

2. Catalase test

(1) Aseptically pick up some bacterial cells from the colony or slant growth with a sterile wooden applicator stick or toothpick.

(2) Deposit the mass of bacterial cells adhering to the stick or toothpick onto a clean glass microscope slide.

(3) Pipette a drop of hydrogen peroxide onto the mass of bacterial cells adhering to the loop.

(4) Observe the drop of hydrogen peroxide to see if bubbles evolve. The production of gaseous bubbles indicates the presence of catalase. It is sometimes difficult to see the bubbles when they are small. In these cases a binocular (dissecting) stereomicroscope is used.

(5) Record your results.

Results and Discussions

Record the results on the data sheet.

(1) H_2S production test:

Bacteria	Result	Principle
Salmonella paratyphi A		
A*Salmonella paratyphi B*		

(2) Catalase test:

Bacteria	Result	Principle
B. subtilis		
Clostridium sp.		

实验二十六　淀粉酶和脲酶试验

【实验目的】

1. 掌握微生物鉴定中常用的几种酶催化的反应原理及结果判断方法。
2. 熟悉微生物生化反应中各种培养基的设计和用途。

【实验原理】

淀粉酶催化淀粉分解为麦芽糖，使得微生物可以吸收利用之。
脲酶催化尿素的分解，微生物在代谢过程产生的毒性废物可以借此消除。

【实验材料】

1. 细菌

大肠埃希菌培养物；枯草杆菌；普通变形杆菌培养物。

2. 试剂

含有 0.2% 淀粉的营养琼脂平板；Lugol 碘液；Christensen 尿素斜面。

3. 其他

接种环；灭菌牙签；灭菌枪头；载玻片等。

【实验方法】

1. 淀粉酶试验

（1）按照无菌操作，将试验细菌点种于淀粉营养琼脂平板中央。

（2）置 37℃ 恒温箱中培养 24 ~ 48 小时。

（3）培养结束后，往平板中加入碘液，倾去多余碘液，观察结果。蓝色部分表示存在淀粉和碘液的反应，透明区表示有淀粉酶将淀粉分解。

（4）记录实验结果。

2. 脲酶试验

（1）将试验菌接种于 Christensen 尿素斜面，另设一支斜面作对照。

（2）置 37℃ 恒温箱中培养 24 ~ 48 小时。

（3）观察斜面上有无紫红色产生，若有，则为脲酶试验阳性。若为黄色，则为脲酶试验阴性。

（4）记录结果。

【结果与讨论】

1. 淀粉酶试验结果

生化反应	细菌	实验现象	结果	原理
淀粉酶试验	大肠埃希菌			
	枯草杆菌			

+为试验阳性；－为试验阴性。

2. 脲酶试验结果

生化反应	菌名	实验现象	结果	原理
脲酶试验	大肠埃希菌			
	普通变形杆菌			

+为试验阳性；－为试验阴性。

26 Amylase and Urease Test

Purpose

1. To understand the principle of enzymes tests and the judgment of these tests.
2. To be acquainted with the design and purpose of test media.

Principle

The extracellular enzyme α - amylase catalyzes the breakdown of starch into maltose. The smaller, more soluble maltose molecules can then enter the cell where they are further metabolized.

The enzyme urease catalyzes the breakdown of urea. Microorganisms that can produce this enzyme are able to detoxify a waste product and to derive metabolic energy from its utilization.

Materials and Apparatus

1. Cultures

Escherichia coli, *Bacillus subtilis*, *Proteus vulgaris.*

2. Reagents

Petri plates containing sterile nutrient agar supplemented with 0.2% soluble starch Lugol's iodine (or Gram's iodine), Christensen urea agar slants or urea broth tubes, Nutrient agar or tryptic soy agar.

3. Apparatus

Sterile wooden applicator sticks or toothpicks, filter paper, pasteur pipettes.

Procedures

1. α - amylase test

(1) Inoculate the plates with each microorganism by making a single streak of the microorganism down the middle of the plate. Label all plates.

(2) Incubate at 37℃ 24 ~ 48h.

(3) After incubation, flood the plate with iodine solution and observe. The development of a blue color indicates the presence of residual starch; areas where starch has been hydrolyzed by amylase will appear clear.

(4) Record you results.

2. Urease test

(1) Inoculate the surface of the Christensen urea agar slants with each microorganism. As an alternative, you may inoculate urea broth tubes with each microorganism. Leave one slant (or broth) as a control. Label the tubes.

(2) Incubate at 37℃ for 24 ~ 48h.

(3) After incubation, observe the tubes. The appearance of a red – violet color indicates a positive test; negative tests are yellow – orange.

(4) Record your results.

Results and Discussion

Record the results on the data sheet.

(1) α – amylase test

Bacteria	Result	Principle
E. coli		
S. subtilis		

(2) Urease test

Bacteria	Result	Principle
E. coli		
P. vulgaris		

实验二十七　硝酸盐还原试验

硝酸盐还原过程与某些微生物的厌氧呼吸偶联。根据微生物还原硝酸盐能力的不同可以对其进行鉴定，如 *Neisseriaceae* 属、*Neisseria mucosa*，*Moraxella catarrhalis* 和 *Kingella denitrifican* 均具有还原硝酸盐的能力。*M. catarrhalis* 和 *K. denitrificans* 往往会误判为 *N. gonorrhoeae*，他们的区别就在于前者硝酸盐还原试验为阳性，而后者为阴性。

【实验目的】

了解微生物将硝酸盐还原为亚硝酸盐或继续分解亚硝酸盐的能力。

【实验原理】

硝酸盐还原试验基于测定接种后培养基中亚硝酸盐浓度的变化，若有亚硝酸盐产生，加入对氨基苯磺酸试剂（sulfanilic acid，试剂 A）后，可产生无色的亚硝酸盐 - 对氨基苯磺酸复合物，当加入 α - 萘胺试剂（α - naphthylamine，试剂 B）时，则可产生樱桃红色的百浪多息沉淀。

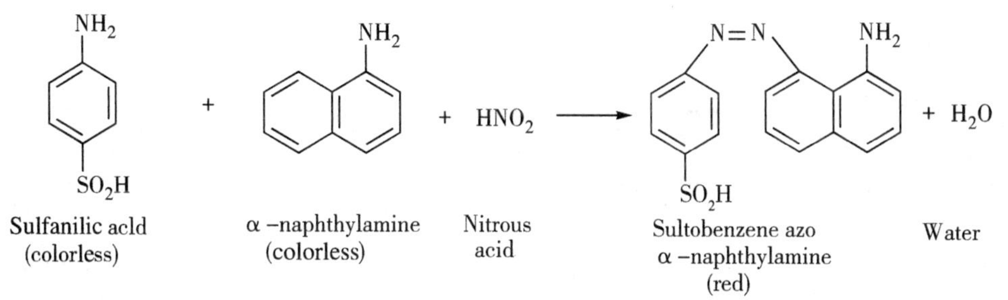

【实验材料】

1. 菌种

大肠埃希菌，粪产碱菌，铜绿假单胞菌。

2. 培养基

胰蛋白胨硝酸盐肉汤。

3. 试剂

溶液 A（对氨基苯磺酸）、溶液 B（α - 萘胺）、锌粉。

4. 器材

酒精灯、接种环。

【实验方法】

1. 将每支胰蛋白胨硝酸盐肉汤分别接种试验菌株，培养48小时

2. 滴加10~15滴试剂A和试剂B，静置5分钟，观察是否有红色产生。

3. 若无颜色变化，可加入少许锌粉，培养基保持无色状态，表明产生了硝酸盐和亚硝酸盐；如果培养基变为红色，表明无硝酸盐还原作用。

【结果与讨论】

1. 先加5滴试剂A溶液，再加入5滴B试剂于各培养基中，观察是否有红色生成，并将结果记录于表格中。

2. 未出现红色的培养基，则再加入微量锌粉，观察有无红色形成，并将结果记录于表格中。

菌株	结果	原理
大肠埃希菌		
铜绿假单胞菌		
粪产碱菌		

27 Nitrate Reduction Test

The reduction of nitrate may be coupled to anaerobic respiration in some species. Bacterial species may be differentiated on the basis of their ability to reduce nitrate to nitrite or nitrogenous gases. Among the *Neisseriaceae* of human origin, strains of *Neisseria mucosa*, *Moraxella catarrhalis*, and *Kingella denitrificans* reduce nitrate. Strains of *M. catarrhalis* and *K. denitrificans* have been misidentified as N. gonorrhoeae. The nitrate reduction test permits differentiation between these species which are nitrate positive and *N. gonorrhoeae* (nitrate – negative).

Purpose

To understand the ability to reduce nitrate to nitrite of some bacteria.

Principle

The nitrate reduction test is based on the detection of nitrite in the medium after incubation with an organism. If present in the medium, nitrite will react with sulfanilic acid (Nitrate reagent A) to form a colorless complex (nitrite – sulfanilic acid). This complex will then yield a red precipitate (prontosil) when nitrate reagent B (alpha – naphthylamine) is then added to the test.

Materials and Apparatus

1. Cultures

Escherichia coli, *Alcaligenes faecalis*, *Pseudomonas aeruginosa.*

2. Reagents

Media: Trypticase nitrate broth, Sulfanilic acid solution (Nitrate reagent A), Alpha – Naphthylamine solution (Nitrate reagent B), Zinc powder.

3. Apparatus

Inoculating loop, Bunsen burner.

Procedures

1. Inoculate nitrate broth with an isolate and incubate for 48h.

2. Add 10 ~ 15 drops each of sulfanilic acid and *N*, *N* – dimethyl – 1 – naphthylamine. If the bacterium produces nitrate reductase, the broth will turn a deep red within

5min at this step.

3. If no color change is observed, then the result is inconclusive. Add a small amount of zinc to the broth. If the solution remains colorless, then both nitrate and nitrite reductase are present. If the solution turns red, nitrate reductase is not present.

Results and Discussions

Record the results on the data sheet.

Bacteria	Result	Principle
E. coli		
P. aeruginosa		
A. faecalis		

实验二十八　石蕊牛乳试验

【实验目的】

利用石蕊试剂检查微生物在脱脂牛奶中的反应情况。

【实验原理】

脱脂牛奶的主要成分是酪蛋白，试验培养基中含有乳糖及矿物质。细菌可利用脱脂牛奶作为营养物质，从而产生不同的变化。石蕊是常用的酸碱指示剂，可用于检查微生物对牛奶的利用情况。

【实验材料】

1. 菌种

大肠埃希菌，普通变形杆菌，黏性产碱菌。

2. 培养基

石蕊牛乳培养基。

3. 器材

酒精灯、接种环。

【实验方法】

1. 标记 3 种细菌培养物。
2. 分别接种于石蕊牛乳培养基，放置培养。
3. 48 小时及 5 天后记录反应情况。

【结果与讨论】

记录试验结果于下表。

菌株	产酸	产碱	胨化	凝乳	还原	产气	黏胶状
大肠埃希菌							
变形杆菌							
黏性产碱菌							

28 Litmus Milk Test

Purpose

To differentiate bacteria based on various reactions that occur in skim milk suplemented with a litmus pH indicator.

Principle

Milk is a complex nutritional source that contains proteins (mainly casein) in an aqueous solution of lactose and minerals. Bacterial enzymes alter the media and may bring about various changes. Litmus is added to the medium to detect pH changes that may occur as a result of these enzymatic reactions. Above a pH of 8. 3 litmus is blue, while below a pH of 4. 5 litmus is red.

Materials and Apparatus

1. Cultures

Escherichia coli, *Proteus vulgaris*, *Alcaligenes viscoelastic*.

2. Reagents

Litmus milk medium.

3. Apparatus

Inoculating loop, Bunsen burner

Procedures

1. Label the litmus milk tube with the name of 3 cultures.
2. Use the aseptic technique, inoculate the tube of litmus milk and place them in the designated area.
3. Examine the tubes after 48 hours and again after 5 days, note the reactions.

Results

Record the results the sheet, using a plus or minus in each case.

strain	Acid reaction	Basic reaction	Peptonization	Curd	Reduction	Gas	Ropiness
E. coli							
P. vulgaris							
A. viscoelastic							

第五部分　病毒学实验

实验二十九　流感病毒的培养与检测

【实验目的】

1. 学会采用鸡胚接种法培养流感病毒。
2. 了解检测流感病毒的血凝试验法。

【实验原理】

鸡胚培养为病毒常用的培养方法之一，该法操作简便、成本低廉，可用于黏病毒、痘病毒、疱疹病毒、脑炎病毒等多种病毒的培养。鸡胚有 4 个部位可用于接种，即绒毛尿囊膜、尿囊腔、羊膜腔和卵黄囊，如图 29－1 所示。流感病毒的培养常采用尿囊腔接种法或羊膜腔接种法（本实验以尿囊腔接种法为例）。

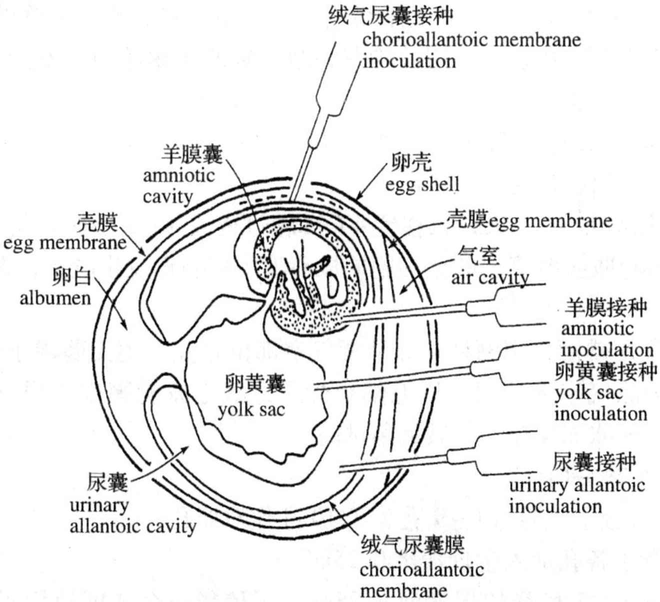

图 29－1　鸡胚的结构和接种途径

Fig. 29－1 Schematic structure of embryonated chicken & inoculating route

血凝试验是检测流感病毒的常用方法。由于流感病毒包膜上有血凝素（HA）分子，能与禽类及多种哺乳动物红细胞膜上的血凝素受体结合，使得红细胞凝集，故称为血凝试验。通过血凝试验可测定血凝效价，从而反映尿囊液中流感病毒的滴度。

【实验材料】

1. 病毒与鸡胚

流感病毒或其他黏病毒（10^{-8}稀释）溶液、10~12 日龄鸡胚。

2. 试剂

无菌石蜡、无菌生理盐水、0.5%鸡红细胞悬液。

3. 其他

蛋座、检卵灯、镊子、碘酒和酒精棉球、砂轮、1ml 的无菌注射器、6 号针头、无菌试管、20 孔反应板、刻度吸管等。

【实验方法】

1. 尿囊腔接种

（1）准备鸡胚。孵育前的鸡卵先用清水以毛刷清洗，再用干布擦干，放入孵箱内进行孵育（37℃，相对湿度 45%~60%），孵育 3 日后，鸡卵每日翻动 1~2 次。孵至第 4 日，用检卵灯观察鸡胚发育情况。未受精卵，只见模糊的卵黄黑影，不见鸡胚的形迹，这些鸡卵剔出不用。活胚可看到清晰的血管和鸡胚的暗影，有些还可以看见胚动，随后每日观察一次，胚动呆滞、没有运动的或血管昏暗模糊者，即可能已死或将死的鸡胚，要随时加以淘汰。

（2）在选取的 10~12 日龄活鸡胚上用铅笔画出气室、鸡胚及大血管的位置，先用碘酒、再用酒精棉球消毒气室部蛋壳，用砂轮在胚胎位置的尿囊腔与气室交界处磨一小口，定为注射入口，勿损伤蛋膜。

（3）用注射器抽取病毒，在注射口使针头与卵壳呈 30°角插入 0.5cm 深度，注入 0.1~0.2ml，注射完后用熔化的石蜡封口。

（4）置 35℃孵箱培养 3~4 天，再移置 4℃冰箱过夜或 4~6 小时。将气室朝上放置，使鸡胚血液凝固，避免收获时流出的血细胞同尿囊液中的病毒发生凝集，造成病毒滴度的下降。

（5）收获时，先用碘酒再用酒精棉球消毒气室部位蛋壳，用无菌镊子除去气室蛋壳，再用无菌眼科镊子去除气室部壳膜，用毛细吸管通过绒毛尿囊膜进入尿囊腔，吸取尿囊液，放入小瓶或试管内冰冻保存，并做无菌试验。

2. 血凝试验（表 29-1）

（1）在 20 孔反应板上编号（1~9 孔，第 10 孔作对照孔）。

（2）用 1ml 吸管于各孔加入生理盐水 0.25ml。

（3）第 1 孔加入 1∶5 稀释的尿囊液 0.25ml，每稀释一个浓度换用吸管，稀释后取 0.25ml 放入第 2 孔中，依次稀释到第 9 孔为止，最后从第 9 孔中弃去 0.25ml。

（4）在各孔加入 0.5%鸡血细胞悬液 0.25ml，摇匀，室温下静置 30~60 分钟后，观察结果。

表 29－1　血凝试验的操作程序

孔号	1	2	3	4	5	6	7	8	9	10
生理盐水（ml）	0.25	0.25	0.25	0.25	0.25	0.25	0.25	0.25	0.25	0.25
尿囊液（ml）	0.25	↘	↘	↘	↘	↘	↘	↘	↘	↘弃
初始病毒稀释度	1∶10	1∶20	1∶40	1∶80	1∶160	1∶320	1∶640	1∶1280	1∶2560	去 0.25
0.5% 鸡红细胞（ml）	0.25	0.25	0.25	0.25	0.25	0.25	0.25	0.25	0.25	0.25
最终病毒稀释度	1∶20	1∶40	1∶80	1∶160	1∶320	1∶640	1∶1280	1∶2560	1∶5120	－

（5）结果判断见表 29－2。

表 29－2　血凝试验结果判断标准

孔底所见	结果
凝集的血细胞颗粒在底部铺开成网状，无血细胞下滑，即 100% 血细胞凝集	＋＋＋＋
基本同上，但边缘不齐整，絮状凝集略有下沉趋向，75% 血细胞凝集	＋＋＋
孔底中央血细胞呈一环状，四周有絮状小块，50% 以上血细胞凝集	＋＋
血细胞于孔底呈一小团，边缘不光滑，四周有小凝集块，25% 以上血细胞凝集	＋
血细胞于孔底呈一小圆点，边缘光滑整齐，血细胞无凝集现象	－

出现“＋＋”（50%）血球凝集现象者，其最高稀释度即为流感病毒的血凝效价。如图 29－2 所示，A4 孔即为血凝效价判定点，因此其最终稀释度 1∶160 为血凝效价。

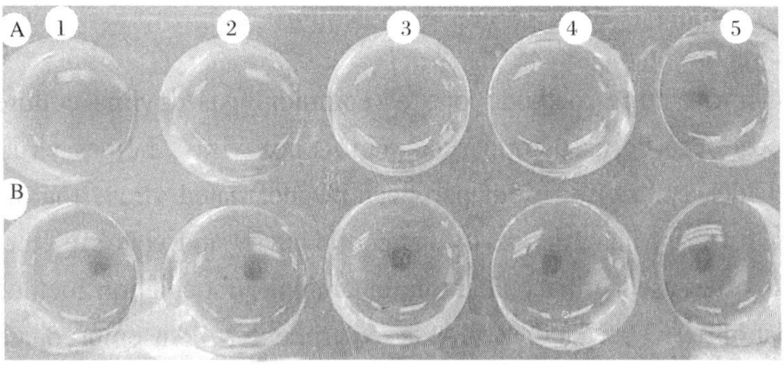

图 29－2　血凝试验的结果

Fig. 29－2 Schematic results of hemagglutination test

【结果和讨论】

1. 血凝实验的原理是什么？

2. 血凝试验结果记录：

孔号	A1	A2	A3	A4	A5
结果					
孔号	B1	B2	B3	B4	B5
结果					

3. 鸡胚尿囊液中有无流感病毒增殖？病毒的血凝效价是多少？

PART FIVE Experiments for Virology

29 Culture and Detection of Influenza Virus

Purpose

1. To culture influenza virus by chick embryo inoculation.
2. To learn the hemagglutination test using for influenza virus detection.

Principle

Chick embryo culture method, one of commonly used virus culture methods, is simple, low cost, can be used for visco – virus, pox virus, herpes virus, encephalitis virus, *et al.* Chick embryo has four parts can be inoculated virus, that is, chorioallantoic membrane, urinary allantoic cavity, amniotic cavity and yolk sac, the cultivation of influenza virus often used urinary allantoic or amniotic inoculation method (urinary allantoic inoculation is to be adopted in this experiment). An embryonated egg is diagrammed in Fig. 29 – 1 as it would appear at 10 ~ 12 days showing the important structures involved in virus culture. The time of injection and area into which the embryo is inoculated depend on the virus.

Visco – viruses (including influenza viruse) have been shown to possess the capacity of adsorbing to red blood cells and in many cases causing agglutination of these cells. It is called the haemagglutinin test. This haemagglutination (HA) test furnishes a relatively simple, quick, convenient and fairly quantitative way of detecting, identifying and titrating these viruses.

Materials and Apparatus

1. Virus and chick embryo: the influenza virus or other visco – virus (10^{-8} dilution) solution, 10 ~ 12 – day – old chick embryo. 2. Reagents: sterile paraffin, sterile saline, 0.5% chicken red blood cell suspension. 3. Others: egg holders, eggs seized lights, tweezers, iodine and alcohol cotton balls, grinding wheel, 1ml sterile syringes,

6 – gauge needles, sterile test – tube, 20 – hole reaction plate, calibrated pipette.

Procedures

1. Chick embryo inoculation (**urine allantoic cavity**)

(1) Preparation of eggs: it is necessary that the fertile eggs be incubated at 37℃ for 10 ~ 12 days in a humid atmosphere (relative humidity: 45% ~60%). Also, it is required that eggs be turned once a day proper embryo development.

(2) After the desired incubation period candle the egg, determine the viability of the embryo (does it move?) and the position of the air sac.

(3) Place the egg on a firm surface. Swab a small area with 70% alcohol immediately below the air sac. Then gently puncture the eggshell with a punch or drill. Be careful not to penetrate the chorioallantioic membrane (CAM). Also, swab another area over the air sac again puncture the shell.

(4) In order to obtain virus for subsequent work, and attempt to obtain an end-point of infection, dilution of influenza virus will be inoculated, thus dilutions of influenza virus will be provide.

(5) Draw 0. 2ml of virus solution into a 1ml syringe fitted with a 6 – gauge needle. Expel any air bubbles.

(6) Carefully insert the needle through the hole in the shell below the air sac and through the CAM about 0. 5 cm, directing it slightly toward the air sac.

(7) Inject exactly 0. 1 ~0. 2 ml by twisting the barrel of the syringe inward.

(8) Carefully remove the needle and seal the holes with small pieces of plastic tape, label each egg.

(9) Place the egg at 35℃ and incubate for 72 ~96h.

(10) Retrieve your eggs from the incubator. Chill by placing them in the refrigerator overnight (or for at least 4 ~6h before step 11).

(11) Place the egg in a holder contained in a porcelain pan.

(12) Break the shell over the air sac by penetrating it with the point of a pair of forceps. Carefully peel black the shell over the air sac.

(13) Tear opens the shell membrane and allantois. Insert a Pasteur pipette equipped with a rubber bulb into the allantoic cavity, being careful not to rupture any blood vessels.

(14) Aspirate 2 ~3ml of fluid and transfer to a sterile tube.

(15) Empty embryo and remaining allantoic fluid into a petri dish and further tear the allantoic membrane to encourage bleeding. Allow to sitting at room temperature for 10 minutes then observing for clumping of erythrocytes. If virus is present a positive test may be observed.

(16) Discard the embryo into appropriate container.

2. Hemagglutination (HA) test

(1) To the plastic microtitre plate, add 250μl of PBS to each of the wells in the first two rows of the plate (A1 – A5 and B1 – B5). Use the 10th well of B5 for the negative controls (no allantoic fluid).

(2) Add 250μl of 5 – fold diluted influenza – containing allantoic fluid (AF) to A1 and mix gently. With a fresh pipette tip, transfer 250μl to A2 and mix gently. Repeat this procedure for the whole row, discarding 250μl from the last well in the series (B1 – B4).

(3) Using a clean pipette tip, add 250μl of RBCs to each well, including the negative controls (resuspend the cells before using). Work from right to left (starting with the negative controls and finishing with the lowest virus dilution).

(4) Mix the cell suspensions by tapping the plate gently and incubate at room temperature for 30 ~ 60 min.

(5) Read and record the titration

In wells where enough influenza virus is present, all the RBCs have agglutinated (i. e. are linked together in a 3 – dimensional matrix, see Fig. 29 – 2). In wells where the concentration of influenza virus has been diluted below a critical point (1 HAU), the cells settle out at the bottom of the well to form a 'button'.

The minimum amount of virus which will cause about 50% of the RBCs hemagglutinating in a well is known as one hemagglutinating unit (HAU). Then the virus titre in the allantoic fluid of chicken embryo is 1 : 160.

Results and Discussion

1. What is the principle of hemagglutination test?

2. Results of hemagglutination test :

No. of the hole	A1	A2	A3	A4	A5
result					
No. of the hole	B1	B2	B3	B4	B5
result					

3. Whether or not influenza virus is proliferating in chicken embryo allantoic fluid? And what is are the hemagglutination titer of the virus?

实验三十　噬菌体分离和效价测定

【实验目的】

1. 富集人类肠道菌中的噬菌体；
2. 从污水中分离相关的噬菌体；
3. 了解噬菌体效价测定的基本方法。

【实验原理】

病毒是一类非细胞型的微生物，仅由核酸和蛋白质组成。病毒必须侵入到宿主细胞内，才能够得以复制，属于专性细胞内寄生物。常见的病毒有植物病毒、动物病毒和微生物病毒。其中以细菌为宿主的病毒称为噬菌体。噬菌体完成复制循环后往往导致宿主细胞的裂解（图 30－1），此类噬菌体称为烈性噬菌体。

在自然环境中，噬菌体的滴度往往很低，因为环境中宿主细胞的数量远远低于实验室中纯培养中细菌的数量。因而在分离和研究噬菌体之前，必须进行富集浓缩。该过程即将含有噬菌体的样品接种于肉汤培养基中的宿主细菌细胞，噬菌体侵染宿主细胞，并且在细胞内增殖，除去未裂解的细胞以及细胞碎片，培养液即可用于噬菌体活性的测定。

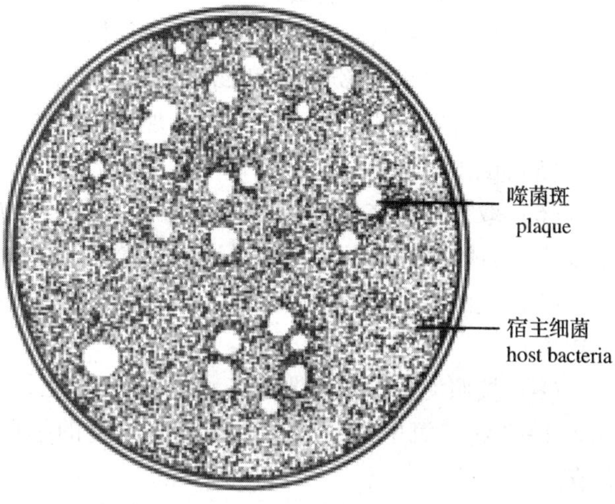

图 30－1　琼脂平板上的噬菌斑

Fig. 30－1 Plaques on the plate

为了分析噬菌体的感染性，可制备固体培养基平板，接种细菌使之可以形成菌苔。噬菌体悬液经稀释后涂布在菌苔表面，若无烈性噬菌体，固体平板表面形成连片的菌苔；若有烈性噬菌体存在，将形成透亮的区域，称为噬菌斑。每个噬菌斑表示了由一个噬菌体裂解触发的裂解循环。噬菌体的效价就是 1 毫升培养液中所含噬菌体的数量。

【实验器材】

1. 细菌

大肠杆菌培养物。

2. 试剂

10×酪蛋白大豆液体培养基（TSB）、酪蛋白大豆琼脂（TSA）、TSA 半固体高层(5ml/管)、污水样品、乙醇、氯仿。

3. 其他

250ml 三角瓶、膜过滤装置（0.45μm）、镊子、50ml 离心管、离心机、灭菌枪头、移液管、旋涡混合器；涂布棒、灭菌小试管 5 支，灭菌 1 ml 吸管 10 支，水浴锅等。

4. 细菌和噬菌体

大肠杆菌 18h 培养液，大肠杆菌噬菌体 10^{-2} 稀释液（用肉汤蛋白胨液体培养基稀释）。

5. 试剂和培养基

肉汤蛋白胨琼脂平板（10 ml 培养基，2% 琼脂，做底层平板用）5 个，含 4ml 琼脂培养基的试管（0.7% 琼脂，做上层平板用）5 管。

【实验方法】

1. 噬菌体分离

（1）于 250ml 三角瓶中加入 5ml 大肠杆菌培养物、10×TSB 以及 45ml 污水样品（注意：污水中可能含有致病菌，应戴上手套和用移液管转移污水样品）。

（2）将三角瓶置于 37℃ 培养 18~24 小时。

（3）将富集物转移到 50ml 离心管，5000rpm 离心 10 分钟。

（4）弃去颗粒成分，取上清部分加到另一洁净离心管，5000rpm 离心 10 分钟。

（5）倒出上清部分，不要带动沉淀部分。

（6）取出浸泡在 75% 酒精中的镊子，火焰中灭菌，冷却后夹取已灭菌的滤膜，放在细菌滤器的烧结滤板上，装配好过滤装置。

（7）过滤离心上清，若还残留细菌细胞及碎片，过滤将进行得很慢，此时须再进行一次离心。

（8）按照无菌操作的要求，将滤液转移到无菌三角瓶，如不立即实验，可在滤液中加入 2~3 滴三氯甲烷，用于杀死随后生长的细菌，但对噬菌体无作用，使滤液可长期保存。

（9）标记好半固体高层，在水浴中熔化，用于噬菌体的扩增。另标记一管作为对照；

（10）半固体高层冷却到 50℃，加入 0.5ml 测试用菌的过夜培养物。

（11）无菌操作加入噬菌体滤液 1 滴，不加滤液的为对照（若有必要，须将滤液作适当稀释）。

（12）混合，迅速倾注在 TSA 平板上，在平板形成之前避免凝固块的产生，凝固块会干扰噬菌斑的判断。

（13）待上层 TSA 凝固后，将平板置于 37℃ 倒置培养 18~24 小时，观察有无噬菌斑产生。

（14）用接种针挑取单个噬菌斑，转入蛋白胨水中进行稀释并如上述方法制备噬菌斑，如此反复三次，达到纯化的目的。

2. 噬菌体效价的测定

（1）取五支装有半固体高层的试管，做好标记，在水浴上熔化管内培养基。

（2）待管内培养基温度降至50℃，加入0.5ml过夜培养的检测用菌。

（3）无菌操作接入稀释的噬菌体悬液0.1ml。

（4）混匀试管，迅速倾注半固体高层于TSA平板。

（5）待上层半固体培养基凝固，置于37℃ 培养箱培养18～24小时。

（6）观察并记录结果。

【结果和讨论】

1. 噬菌体分离

（1）试绘图表示平板上出现的噬菌斑。

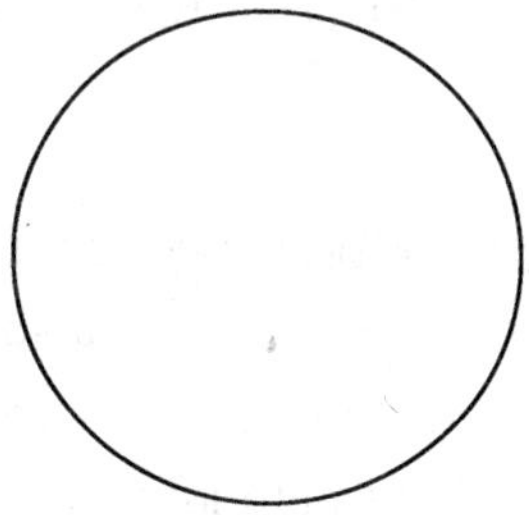

（2）为什么噬菌体对一种大肠杆菌敏感而对另一种大肠杆菌却不敏感？

（3）为什么不是所有菌苔上的细菌都被噬菌体裂解？

（4）为什么污水中的细菌不被其周围的病毒完全杀死？

2. 噬菌体效价的测定

（1）将平板中各稀释度的噬菌斑记录于下表，并按下式计算噬菌体的效价。

效价（单位/ml）＝平均噬菌斑数×稀释度×取样量折算数

噬菌体稀释度	1	2	3	4	5
噬菌斑形成					
噬菌体效价					

（2）噬菌体效价测定时应注意哪些操作？

30 Isolation and Titer Detection of Bacteriophage

Purpose

1. To enrich for bacteriophage that infect bacteria found in the human intestinal tract;

2. To isolate bacteriophages from sewage;

3. To understand the method of detecting bacteriophage titer.

Principle

Viruses are a kind of cellular microorganisms composed of one kind of nucleic acid surrounded by a protein coat. Viruses must invade a specific host cell, in which they utilize the cellular machinery to reproduce more virus particles. Thus they are referred to as obligate intracellular parasites. Viruses are known to infect plants, animals, and bacterial cells.

Viruses which infect and replicate within specific bacterial host cells are known as bacteriophage, or simply phage. The normal replication cycle of bacteriophage results and discussions in the lysis of the host cell at the completion of the viral replication cycle; thus, these bacteriophage are referred to as lytic phage.

In natural environments, the phage titer is usually low because host numbers are far less than those encountered in laboratory pure cultures. Therefore, before the phage can be isolated and studied, its numbers must be increased by an enrichment procedures. This can be achieved by incubating samples containing the phage with a suitable host microorganism in a broth culture. phage present will infect the host bacterium, thereby increasing their numbers in the culture. After any unlysed cells and cell debris have been removed, the broth culture can be used to test for phage activity by looking for plaque formation in the bacterium used for enrichment. However, although the indicator strain may be lysed, the phage may not infect all strains of the bacterium. In some cases, this may be due to differences in the make - up of cell wall between the different strains, which are important in viral recognition.

To assay for infective bacteriophage, a lawn of bacteria is prepared on a suitable solid nutrient medium, and dilutions of the phage suspension are then spread over the same surface. In the absence of lytic bacteriophage, the bacteria form a confluent lawn of growth. Lysis by bacteriophage is indicated by the formation of a zone of clearing or plaque within the lawn of bacteria. Each plaque corresponds to the site where a single bacteriophage initiated its lytic reproductive cycle. The titer of phage means the number of phage in 1ml of lytic filtrate.

Materials and Apparatus

1. Cultures

Cultures of *Escherichia coli*.

2. Regent

10 × strength trypticase soy broth, trypticase soy agar plates, raw sewage, ethanol, chloroform.

3. Apparatus

125ml Erlenmryer flasks, Sterile membrane filter assemblies (0.45μm filters), Forceps, 50 ml centrifuge tubes, Centrifuge, Soft agar tubes (TSA: 5 ml/tube), Sterile pipettes, Graduated cylinders, Water bath at 50℃, Vortex mixer.

Procedures

1. Isolation of Bacteriophage

(1) Inoculate a 125ml flask with 5ml of bacterial culture, 5ml of 10 × strength TSB, and 45ml of raw sewage. Use *Escherichia coli* and inoculate one culture per flask. (Note: Raw sewage may contain potential pathogens; wear gloves and use the graduated cylinders to deliver the raw sewage).

(2) Incubate the flask for 18 ~ 24h at 37℃ with aeration.

(3) Deliver the enrichment mixture to 50ml centrifuge tubes, filling them to the same level. Centrifuge the tubes at 5000 rpm for 10 min.

(4) Carefully pour off the supernatant into clean centrifuge tubes, leaving behind the pellet. Repeat the centrifugation step.

(5) Decant the supernatant, being careful not to disturb the pellet.

(6) Assemble the filter apparatus: Flame the forceps after dipping them in alcohol. Aseptically remove a 0.45μm membrane filter from its sterile package and place it on sterile glass filter base which is positioned in a vacuum flask. Place the filter – funnel over the membrane filter and clamp it on with the filter clamp.

(7) Filter the centrifuged supernatants by vacuum filtration through the 0.45μm filter. If the centrifugation step effectively removes all the bacteria cells and any cell debris remaining in the broth, the filtration will proceed relatively rapidly. If large numbers of bacteria and debris remain, the filter will clog. If this occurs, it may be necessary to include an additional centrifugation step.

(8) Aseptically transfer the filtrate to a sterile flask. This may be stored for future use by adding 2 ~ 3 drops of chloroform to the broth (Caution: Remember that chloroform is highly flammable). The latter will kill any bacteria that attempt to grow, but it will not harm the phage.

2. Titer Detection of Bacteriophage

(1) Label five tubes of soft agar 1, 2, 3, 4, and 5. Melt them in boiling water. Soft agar contains less agar (0.7%) than media used for most bacteriological applications. It promotes better development of plaques formed by the bacteriophage. Also label five TSA plates to correspond to your tubes.

(2) To each tube of soft agar, which has been cooled to 50℃, add 0.5 ml of an overnight culture of the bacterial strain to be tested. You will need to set up a series of tubes for the bacterial strain that you used for the enrichment step, plus tubes for any other strains of this microorganism that you test. Hold the tubes of soft agar at 50℃ until you are ready to pour them.

(3) Aseptically add the following amounts of filtrate to the tubes: tube 1, one drop; tube 2, two drops; tube 3, four drops; tube 4, eight drops. Tube 5 receives no filtrate because it is a growth control. (Note: It may be necessary to dilute the filtrates before spotting due to variations in viral titers from one area to another.)

(4) Cap the tubes and mix by vortexing. Rapidly and aseptically overlay tube 1 onto the trypticase soy agar base in plate 1. Avoid allowing the agar to solidify before pouring, as "lumps" in the agar surface will obscure any plaques that have formed. Repeat this for each tube until you have overlaid all the five plates with their respective tubes of soft agar. The agar base in the plates provides a base for a "lawn" of confluent bacteria on the surface of the plate. Any bacteriophage in the filtrate will infect their respective host, resulting in the bacterial lawn where the host cells have been lysed by the virus.

(5) When the soft agar plates have cooled, invert the plates and incubate them at 37℃ for 18 ~ 24h. If possible, check the plates at 6 ~ 8h for plaque formation.

(6) Record your results and discussions.

Results and Discussion

1. Isolation of Bacteriophage

(1) Record the number of plaque on the plate.

(2) Why would one strain of *E. coli* be sensitive to a particular bacteriophage but another strain be insensitive to the same phage?

(3) What reasons can you give for the fact not all of the cells in the bacterial lawn are lysed by the virus?

(4) Why aren't the bacteria which occur in raw sewage completely killed by their viruses that also occur in this environment?

2. Titer Detection of Bacteriophage

(1) Record the number of plaque on the series plates.

(2) Which steps you should pay attention to in the test?

实验三十一　噬菌体的一步生长曲线

【实验目的】

了解噬菌体一步生长曲线测定的基本方法。

【实验原理】

噬菌体的生长曲线可以通过计算不同时间菌悬液中感染噬菌体颗粒的数量来确定。将细菌加入适当的固体培养基，倾注平板，再涂布经过适当稀释的噬菌体悬液，经过培养，在菌苔处将形成噬菌斑，一个噬菌斑对应一个噬菌体颗粒。

【实验器材】

1. 细菌和噬菌体

大肠埃希菌培养液（5×10^{8}CFU/ml），噬菌体悬液（2×10^{8}phage/ml）。

2. 试剂和培养基

半固体营养琼脂高层（3ml），营养琼脂平板，营养肉汤培养（9.9 ml/管），营养肉汤培养（9.0ml/管）。

3. 其他

涂布棒，灭菌小试管，灭菌枪头，50℃水浴锅等。

【实验方法】

1. 将装有半固体高层的试管，做好标记，在水浴上熔化管内培养基，50℃保温。
2. 加入0.1ml过夜培养的检测用大肠杆菌。无菌操作接入稀释的噬菌体悬液2.2ml，记录时间t=0。
3. 37℃孵育，5分钟后，高速离心5分钟，弃上清，沉淀用9.9ml营养肉汤重悬。
4. 将0.1ml悬液加入9.9ml营养肉汤（稀释100倍），以此出发，再作10^{-3}、10^{-4}、10^{-5}稀释。
5. 接种10^{-3}和10^{-4}稀释度的试管。
6. t=20分钟时，各取0.1ml培养液，分别与半固体培养基混匀，涂布于固体平板。
7. 每隔5分钟，重复上述步骤一次，一直到t=40分钟。
8. 平板于37℃培养24~48小时。
9. 检查噬菌斑数，计算每个取样点的噬菌体数。
10. 绘制生长曲线图。

【结果与讨论】

1. 将平板中各稀释度的噬菌斑记录于下表，并计算噬菌体的效价。

时间（min）	20	25	30	35	40
噬菌斑数					
样品中噬菌体数					

1. 为何不能绘制 t_0 至 t_{20} 的曲线图？
2. 为什么噬菌体的生长曲线称为一步生长曲线？

31 One – Step Growth Curve of Bacteriophage

Purpose

To understand the detecting method of one – step growth curve of bacteriophage.

Principle

The growth curve for the replication cycle of a bacteriophage can be determined by inoculating a suspension of host bacterial cells with the phage and assaying for the number of infective phage particles at various time intervals. To assay for infective bacteriophage, a lawn of bacteria is prepared on a suitable solid nutrient medium, and dilutions of the phage suspension are then spread over the same surface. In the absence of lytic bacteriophage, the bacteria form a confluent lawn of growth. Lysis by bacteriophage is indicated by the formation of a zone of clearing or plaque within the lawn of bacteria. Each plaque corresponds to the site where a single bacteriophage initiated its lytic reproductive cycle.

The spread of infectious phage from the initially infected bacterial cell to surrounding cells results in the lysis of the bacteria in the vicinity of the initial phage particle; hence, the zone of clearing. The soft agar used in these assays permits diffusion of phage to nearby uninfected cells but does not permit the viruses that are produced to move to remote parts of the plate. Plaques do not continue to spread because when a cell is heavily reinoculated with phage before the time of normal lysis, the lysis of the cell is inhibited. The number of plaques that develop and the appropriate dilution factors can be used to calculate the number of bacteriophages in a sample. The lysis of the bacterial cell releases a large number of phage simultaneously, and consequently, a lyric replication cycle exhibits a one – step growth curve.

Materials and Apparatus

1. Cultures

Cultures of *Escherichia coli* (5×10^8 CFU/ml), suspension of bacteriophage titer (2×10^8 phage/ml) .

2. Regent

10 × strength trypticase soy broth, trypticase soy agar plates, soft agar tubes (TSA: 5ml/tube), semisoft agar in tubes (3ml), nutrient agar plates, nutrient broth

tubes, nutrient broth tubes.

3. Apparatus

250ml Erlenmryer flasks, sterile pipettes, graduated cylinders, water bath at 50℃, vortex mixer, centrifuge.

Procedures

1. Using a sterile pipette, aseptically transfer 0. 1ml of the culture of *E. coli* to each of the agar tubes and return them to the water bath.

2. Add 1ml of the phage suspension to 2. 2ml of *E. coli* B culture. Record the time as t_0.

3. Incubate at 37℃. After 5 min centrifuge the tubes in a clinical centrifuge for 5 min at the highest rpm setting. Discard the supernatant and resuspend the pellet in 9. 9 ml of nutrient broth.

4. Prepare a 100 – fold dilution by taking 0. 1ml and adding it to 9. 9 ml of nutrient broth (10^{-2} dilution). Take the 10^{-2} dilution 1. 0ml to 9. 0ml sterile nutrient broth (10^{-3} dilution) and 0. 1 ml to 9. 9 ml Sterile nutrient broth (10^{-4} dilution). Prepare an additional dilution by taking 1. 0 ml from the 10^{-4} dilution broth and transferring it to 9. 0 ml of sterile nutrient broth (10^{-5} dilution).

5. Incubate the 10^{-3} and 10^{-4} dilutions at 37℃.

6. At t = 20 min, transfer 0. 1 ml from the 10^{-3} and 10^{-4} dilutions to separate tubes of semisoft agar. Mix each tube and pour onto separate agar plates.

7. At 5 – min intervals, until t = 40, transfer 0. 1 ml portions from the 10^{-3} and 10^{-4} dilutions to separate tubes of semisoft agar. Mix each tube and pour onto separate agar plates.

8. Incubate the plate at 37℃ for 24 ~ 48h.

9. After incubation, examine the plates for the appearance of plaques (zones of clearing) in the lawn of confluent bacterial growth. Count the number of plaques seen on each plate. From these counts and the dilution factors, calculate the concentration of bacteriophage at each sampling time.

10. Graph a plot of the concentration of bacteriophage versus time.

Results and Discussion

1. Graph a plot of the concentration of bacteriophage versus time.

Time (min)	20	25	30	35	40
Number of plaque					
Number of phage					

2. Why can't your recover bacteriophage between t_0 and t_{20}?

3. Why is the growth curve for bacteriophage called a single – step growth curve?

第六部分　微生物遗传学实验

实验三十二　微生物的遗传与变异现象

【实验目的】

1. 观察细菌的菌落变异现象。
2. 观察菌体的形态变异现象。

【实验原理】

遗传和变异是微生物的基本特性。细菌的变异现象有很多，包括菌落和细胞形态变异，荚膜、芽孢、鞭毛等结构变异，以及生化变异如毒力和抗药性变异。一些变异只是表型的改变，而一些变异是遗传物质的改变（称为突变）。只有后者可传递给子代。突变可以自发产生或用诱变剂诱发产生。

【实验材料】

1. 菌种

普通变形杆菌 18～24 小时斜面培养物，黏质沙雷菌 18～24 小时斜面培养物，金黄色葡萄球菌 24 小时肉汤培养物，光滑型、粗糙型大肠埃希菌琼脂斜面 18～24 小时培养物，鼠疫杆菌（无毒株）琼脂斜面 18～24 小时培养物。

2. 培养基

普通琼脂培养基，0.1% 石炭酸琼脂培养基，3%～6% NaCl 血琼脂斜面，L 型培养基（牛肉浸液 800ml，蛋白胨 20g，NaCl 50g，琼脂 8g，pH7.4）。高压蒸汽灭菌，等温度降至 50℃左右时，加入无菌人血浆 200ml 后倾注平板。

3. 试剂

青霉素溶液，革兰染液和细胞壁染液。

4. 器材

接种环，低浓度青霉素药物纸片（40μg/片），镊子，涂布棒，无菌平皿，37℃和 25℃培养箱。

【实验方法】

1. 鞭毛变异

普通变形杆菌通常可产生鞭毛，但当某些化学药品存在时，鞭毛生长受到抑制，菌落产生变异。取变异的菌落再分离，去除化学药品影响时，其产鞭毛的特性又得以恢复。

（1）分别在普通琼脂平板和0.1%石炭酸琼脂平板的中央接种普通变形杆菌。

（2）37℃培养24小时后观察菌落有无迁徙现象。

（3）将菌落再分别接种至普通琼脂平板，37℃培养24小时后观察菌落有无迁徙现象。

2. 色素变异

有些细菌产生色素常与培养温度相关，例如黏质沙雷菌在25℃培养时能产生深红色的灵杆菌素而使菌落显色，但在37℃培养时却形成无色菌落。取37℃培养的无色菌落再分离于25℃培养，其产生色素的特性又得以恢复。

（1）菌悬液制备　用约4ml无菌生理盐水刮洗下斜面上的黏质沙雷菌的菌苔，制备成均匀分散的菌悬液。

（2）划线分离　用接种环取满环菌液，在预先制备的普通琼脂平板表面分区划线分离接种，使最后一区出现较多单菌落的分布处。

（3）培养与观察　将接种后的一块平板于25℃培养，另一平板于37℃培养，各培养48小时，观察并记录在不同培养温度下，黏质沙雷菌的菌落产生色素的情况。

（4）复查产色素试验　在37℃培养的平板上挑取不产生红色色素或产生色素不明显的单菌落上的少许菌体，再次划线分离接种于新鲜普通琼脂平板，并于25℃培养48小时。观察能否再现色素分泌的特性。

3. 光滑型与粗糙型菌落变异

细菌菌体抗原称为“O”抗原，细胞壁中脂多糖为群特异性抗原，“O”抗原特异性由脂多糖末端重复结构的多糖残基侧链决定。具有“O”抗原的细菌菌落呈光滑型（S）；丢失多糖侧链的细菌菌落呈粗糙型（R）。在一定条件下，细菌菌落可由光滑型转变为粗糙型，即为S－R变异。平板上光滑型菌落表面光滑、边缘整齐、湿润；粗糙型菌落表面粗糙、边缘不整齐、干皱。S－R变异是一种全面的变异，不仅菌落形态不同，其生化反应能力、抗原性及致病性等也往往发生改变，例如R型菌的致病性明显降低。

（1）分别接种光滑型和粗糙型大肠埃希菌于两个普通平板上。

（2）37℃孵育18～24小时后，观察两型大肠埃希菌的菌落特性。

4. 形态变异

鼠疫杆菌培养在含高盐的培养基中时，可出现大小不等的多形态。

（1）接种鼠疫杆菌（无毒株）于3%～6% NaCl血琼脂斜面，28～30℃培养3日。

（2）取出涂片、革兰染色，与未变异株作对比观察。

5. 细菌L型变异

细菌在有些水解酶（如溶菌酶）或某些抗生素（如青霉素）的作用下，肽聚糖的合成受阻导致细胞壁受损，成为细胞壁缺陷型细菌，称为L型细菌。由于其细胞壁缺陷，在低渗环境中，菌体会裂解死亡，但在高渗含血清的培养基中仍能生长，形成油煎蛋样的细小菌落。

（1）在L型培养基上加入0.05ml金黄色葡萄球菌培养物，然后以涂布棒均匀涂布，待平板稍干后，取青霉素药物纸片1张贴于平板中央，置37℃培养，次日观察有无抑菌圈。

（2）每隔一天用放大镜或在低倍镜下观察抑菌圈内有无油煎蛋样小菌落出现。

（3）选取油煎蛋样菌落和抑菌圈外细菌分别进行涂片，作革兰染色和细胞壁染色后，镜检。抑菌圈内出现油煎蛋样小菌落，经染色后可见细菌呈多形性，经细胞壁染色后，看不清细胞壁结构，与原来的菌对比区别明显。

【结果与讨论】

1. 鞭毛变异实验结果。

细菌名称	培养基	生长状况
普通变形杆菌		

2. 色素变异实验结果。

细菌名称	培养温度	色素产生状况
黏质沙雷菌	25℃	
	37℃	

3. 光滑型与粗糙型菌落变异实验结果。

细菌名称	菌株分型	菌落形态
大肠埃希菌	光滑型	
	粗糙型	

4. 形态变异实验结果。

细菌名称	培养基	菌体形态
鼠疫杆菌（无毒株）		

5. 细菌L型变异实验结果。

细菌名称	培养基	菌落形态	菌体形态
金黄色葡萄球菌			

32 Heredity and Variation of Microorganism

Purpose

1. To be familiar with the phenomena of microorganism's variations of colonies.

2. To be familiar with the phenomena of microorganism's variations of morphology.

Principle

Heredity and variation are common properties of microorganisms. There are many phenomena of bacterial variation including colonial or cellular morphological variation, structural variation of capsule, spore and flagella, biochemical variation such as virulence and drug resistance. Some variations are just phenotypic change and some of them are genetic change called mutation. Only the later can be transferred to progeny.

Mutations occur in one of two ways: ①Spontaneous mutations arise occasionally in all bacteria and develop in the absence of any added agent. ②Induced mutations are the results of the bacteria's exposure to a mutagen, which is a physical or chemical agent.

Materials and Apparatus

1. Strains

18 ~ 24h nutrient agar cultures of *Proteus vulgaris*, *Serratia marcescens*, *Escherichia coli* (S type and R type), *Yersinia pestis* (non - toxic strain), 24h nutrient broth culture of *Staphylococcus aureus*.

2. Media

Nutrient agar, nutrient agar containing 0.1% phenol, 3% ~ 6% NaCl blood agar, L media.

3. Reagents

Penicillin, Gram's - dye, Cell wall dye.

4. Apparatus

Inoculating loop, sterile filter paper discs (containing 40μg penicillin/disc), forceps, spreading rod, sterile petri dishes, 37℃ and 25℃ incubator.

Procedures

1. Flagella Variation

Proteus vulgaris usually have flagella. But the flagella growth was inhibited at presence of certain chemicals and the colony changed. The production of flagella can resume when chemicals removal.

(1) Inoculate *Proteus vulgaris* using inoculating loop in the center of the plates.

(2) Incubate at 37℃ for 24h. Examine the motility of the colonies.

(3) Transfer some material from the colony on nutrient agar containing 0.1% phenol to nutrient agar plate. Incubate at 37℃ for 24h. Examine the motility of the colonies.

2. Pigment Variation

The production of pigment of some bacteria often associated with temperature. *Serratia marcescens* can produce red pigment and form red colonies when cultured at 25℃, but to form colorless colonies when cultured at 37℃. The characteristics of pigment production can resume when cultured at 37℃ again.

(1) Preparation of bacteria suspension Add about 4ml sterile saline to the slant. Wash and scratch the lawn to prepare uniformly dispersed suspension of bacteria.

(2) Streak – plate Transfer a loopful of the bacteria suspension and streak on the plate to let single colonies could be distributed at the last area.

(3) Incubation and Observation Incubate the plate at 25℃ and 37℃ separately for 48h. Observe and record the pigment production of the colonies under different temperature.

(4) Review the production of the pigment Transfer some material from the colorless colonies at 37℃ and streak on nutrient agar plate. Incubate at 25℃ for 48h. Observe the reproduction of pigment.

3. S – R Colony Variation

Bacterial cell antigen is known as the "O" antigen. Cell wall lipopolysaccharide is a group – specific antigen and the specification decided by polysaccharide side chain residues on terminal repeat. The colony was smooth (S) with "O" antigen and rough (R) when loss of polysaccharide side chains. Under certain conditions, bacterial colonies may be smooth (smooth, S) into a rough – type (rough, R) called S – R colony variation. Smooth – type colonies are surface smooth, edge neat and humid. Rough – type colonies are surface rough, edge irregular and dry – wrinkled. S – R variation is a comprehensive variation, not only of different colony morphology, but also biological and chemical response capabilities, antigenicity and pathogenicity, are often changed, such as R – type bacteria decreased pathogenicity.

(1) Inoculate S – type and R – type *E. coli* on the nutrient agar plate separately.

(2) Observe the characteristics of the colonies after 18 ~ 24h incubation.

4. Morphology Variation

(1) *Yersinia pestis* can form many different sizes and morphologies when cultured at high – salt medium. Inoculate *Yersinia pestis* (non – toxic strain) at 3% ~ 6% NaCl blood agar slant and incubate at 28 ~ 30℃ for 3d.

(2) Smear and stain with Gram ' s dyes. Observe the size and morphology and contrast with the normal strain.

5. Bacterial L – type Variation

Bacterial peptidoglycan synthesis will be inhibited and cell wall be damaged under the effect of some hydrolase (such as lysozyme) or certain antibiotics (such as penicillin). The cell wall deficient bacteria, known as L – type, will lyse and die in low – permeability environment due to its cell wall defects. L – type can still grow and form oil – fried egg – like small colonies in hypertonic medium containing serum.

(1) Pipette 0.05ml of the *Staphylococcus aureus* culture onto L – medium plate and spread the culture over the surface of the plate using the sterile glass spreader. Let the plate dry. Place a filter paper disc containing penicillin on the center of the plate. Incubate at 37℃. Observe whether the inhibition zone appear the next day.

(2) Observe whether oil – fried egg – like small colonies appear inside the inhibition zone every other day with a magnifying glass or microscope at low magnification.

(3) Select the oil – fried egg – like colonies and the colonies outside the inhibition zone to smear and stain with Gram's dye and cell wall dye. Observe under microscope. The bacteria from oil – fried egg – like small colonies inside the inhibition zone showed poly – morphological and cell wall not clear compared with the original strain.

Results and Discussion

1. The results of flagella variation.

Strain	Media	Description of growth
Proteus vulgaris		

2. The results of pigment variation.

Strain	Temperature	Pigment
Serratia marcescens	25℃	
	37℃	

3. The results of S – R colony variation.

Strain	Type	Colony
E. coli	S	
	R	

4. The results of morphology variation.

Strain	Media	Morphology
Yersinia pestis (non – toxic strain)		

5. The results of bacterial L – type variation.

Strain	Media	Colony	Morphology
Staphylococcus aureus			

实验三十三　微生物的抗药性变异

【实验目的】

1. 观察细菌的抗药性变异现象。
2. 学习青霉素酶的简易测定方法。
3. 学习用梯度平板法分离抗药性突变株。

【实验原理】

微生物以低频率发生抗药性突变，突变株在含抗生素的培养基上生长，而正常细菌被抑制，因而突变株虽然很少但很容易被检出。抗药突变株的抗药机制包括：①产生改变抗生素结构的酶，例如对青霉素的抗性。②改变细胞膜的通透性，例如对链霉素的抗性。③改变抗生素的作用靶位，例如对链霉素的抗性，链霉素干扰核糖体的翻译过程。

抗药突变株对抗生素具有相对的抵抗力。多剂抗药株显示对几种不同作用机制的药物具有抗药性。

金黄色葡萄球菌青霉素抗药株可产生青霉素酶。该酶水解β-内酰胺环，使青霉素分解为青霉噻唑酸，该酸可被碘氧化，淀粉（遇碘显蓝色）可做为指示剂，即在抗药株的菌落周围因碘被消耗而显出无色透明圈（图 33-1）。

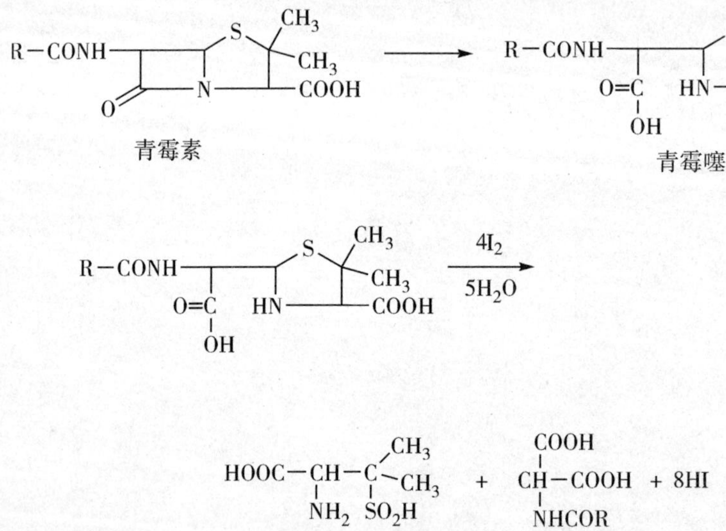

图 33-1　青霉素酶测定原理

Fig. 33-1　Principle for determination of penicilinase

梯度平板法是筛选抗药性突变株的一种有效简便的方法，例如抗链霉素金黄色葡萄球菌突变株的分离，如图 33 - 2 制备双层平板，底层平板倾斜且不含药物，将含药的上层培养基倒在底层上，就会形成药物的含量梯度。将敏感菌涂布在平板上，在高链霉素含量区长出的菌落可能就是抗链霉素突变株。

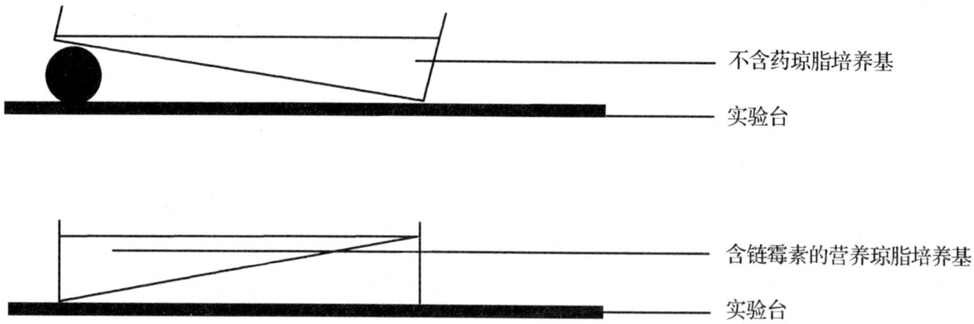

图 33 - 2　梯度平板的制备

Fig. 33 - 2 Preparation of a Gradient - plate

【实验材料】

1. 菌种

金黄色葡萄球菌青霉素敏感株和抗药株的 18 ~ 24 小时斜面培养物，金黄色葡萄球菌 24 小时肉汤培养物。

2. 培养基

普通琼脂培养基。

3. 试剂

青霉素溶液（5U/ml），1ml 青霉素 G 钾盐溶液（20000U/ml），5ml 淀粉琼脂（含淀粉 0.5%，琼脂 0.8%），碘液（0.08mol I_2 溶于 3.2mol/L KI 溶液），链霉素溶液（0.1mg/ml）。

4. 器材

接种环，无菌滤纸片，镊子，涂布棒，无菌平皿，37℃培养箱。

【实验方法】

1. 抗药性变异

（1）将金黄色葡萄球菌青霉素敏感株和抗药株分别密集涂布于琼脂平板表面的一半。

（2）用镊子取滤纸片两片，用青霉素溶液（5 U/ml）浸湿并分别置于（1）平板上。

（3）37℃培养 24 小时，观察滤纸片周围有无抑菌圈形成。

2. 青霉素酶的简易测定法（碘淀粉平板法）

（1）将金黄色葡萄球菌青霉素敏感株和抗药株分别密集涂布于琼脂平板表面的两处（约 1cm²）。

（2）37℃培养 24 小时，敏感株和抗药株生长。

（3）将淀粉琼脂加热融化并恒温于50℃水浴，加入1.0ml青霉素G钾盐溶液(20000U/ml)，混匀后即刻倾注于上述培养好的平板上，铺平待凝。

（4）用滴管吸取碘液铺满平板表面，吸去多余的碘液，倒置于37℃培养箱，10～15分钟后观察抗药菌周围出现无色透明圈。

3. 梯度平板法分离抗药株

（1）融化2支10ml营养琼脂，冷却并维持在约50℃。

（2）将一支铅笔垫在平皿下的一端。无菌操作将融化的营养琼脂倒入平皿覆盖整个底层，待凝成斜面。

（3）平板凝固后，取出铅笔将平板放平。

（4）在第二支融化的营养琼脂试管中加入0.1ml链霉素，用手搓匀，无菌操作将其倒在倾斜的底层平板上，待凝，成链霉素梯度平板。

（5）吸取0.2ml金黄色葡萄球菌培养物于梯度平板，用涂布棒涂布均匀。

（6）37℃倒置培养48小时。

（7）培养后，选择一到两个生长在中间链霉素含量的菌落，向平板上高浓度的方向划线。

（8）37℃倒置培养48小时。

（9）观察划线后的菌落生长情况，在高链霉素含量区域出现的菌落可能是链霉素抗药株。

【结果与讨论】

1. 抗药性变异实验结果。

细菌名称	培养基	生长状况
金黄色葡萄球菌（青霉素敏感株）		
金黄色葡萄球菌（青霉素耐药株）		

2. 青霉素酶测定实验结果。

细菌名称	实验结果
金黄色葡萄球菌（青霉素敏感株）	
金黄色葡萄球菌（青霉素耐药株）	

3. 梯度平板法分离抗药菌实验结果。
4. 细菌抗药性机制是怎样的？
5. 分离抗药突变株为什么要用梯度平板法？
6. 是链霉素导致了抗链霉素突变的产生吗？为什么？
7. 为什么近年来抗药菌在增加？

33 Drug Resistant Variation

Purpose

1. To be familiar with the phenomena of microorganism's variations of drug resistance.

2. To be familiar with a simple method to determine penicillinase.

3. To learn how to isolate a drug resistant mutant by the gradient – plate technique.

Principle

Mutations resistant to antibiotics such as streptomycin occur in a very low ratio in the microbial cell population. However, these rare strains can be easily detected because they grow in the presence of antibiotic concentrations that inhibit the growth of normal bacteria.

In a drug resistant organism, the mutated gene enables the cell to circumvent the anti – microbial effect of the drug by a variety of mechanisms, including: ①The production of an enzyme that alters the chemical structure of the antibiotic, as in penicillin resistance. ②A change in the selective permeability of the cell membrane, as in streptomycin resistance. ③A modification of the target site of antibiotics, as in the resistance to streptomycin, which interferes with the translation process at the ribosomes.

A drug resistant mutant has relative resistance to anti – microbial agents. A multiple drug resistant strain shows resistance to multiple drugs that have different mechanisms.

Penicillin – resistant strain of *Staphylococcus aureus* can produce penicillinase. The enzyme hydrolyzes the β – lactam ring of penicillin to penicilloic acid which can be oxidized by I_2. Starch is used to indicate the presence of I_2 because it can form a blue complex with I_2. I_2 will be exhausted and a clear and colorless area appears around the colony of drug resistant strain.

The gradient – plate is an effective and convenient technique in isolating drug – resistant mutant from a prototrophic (wild type, drug – sensitive) culture. This requires preparation of a double – layer agar plate as illustrated in Fig. 33 – 2. The lower, slanted nutrient agar layer lacks streptomycin. When poured the melted agar containing streptomycin over the lower slanted layer a drug concentration gradient in the surface layer will produce. Following a spread – plate inoculation of the drug – sensitive culture and incubation, the appearance of colonies in a region of high streptomycin concentration is indicative of streptomycin – resistant mutants.

Materials and Apparatus

1. Strains

18 ~ 24h nutrient agar slant cultures of *Staphylococcus aureus* penicillin sensitive strain and penicillin resistant strain, 24h nutrient broth culture of *Staphylococcus aureus.*

2. Media

Nutrient agar.

3. Reagents

Penicillin solution (5U/ml), sterile filter paper discs, starch agar (5ml, starch 0.5%, agar 0.8%), penicillin G potassium solution (20000U/ml), I_2 solution (dissolve 0.08 mol of I_2 into KI solution of 3.2 mol/L), streptomycin solution (0.1mg/ml).

4. Apparatus

Inoculating loop, sterile filter paper discs, forceps, spreading rod, sterile petri dishes, incubator.

Procedures

1. Drug – Resistant Mutation (drug sensitivity test)

(1) Spread the cultures of sensitive and resistant strains to 1/2 of the plate, respectively.

(2) Place two filter paper discs containing penicillin on the area of sensitive and resistant strains, respectively.

(3) Incubate at 37℃ for 24h. Examine the plate for growth.

2. Determination of Penicillinase (I_2 – starch plate method)

(1) Inoculate sensitive and resistant strains on the plate, respectively. The area is about 1 cm^2.

(2) Incubate at 37℃ for 24h.

(3) Melt the starch agar and maintain in 50℃ water bath. Add 1.0ml penicillin G potassium solution (20000U/ml) into starch agar tube, mix quickly and pour on the incubated plate.

(4) After cooling, transfer I_2 solution using pipette on the surface of all the plate. Move away the surplus I_2 solution. Incubate the plate inverted at 37℃ for 10 ~ 15 min. Examine the clear and colorless area around the colony of resistant strain.

3. Isolation of a Streptomycin – Resistant Mutant

(1) Melt the two 10 ml nutrient agar tubes. Cool and maintain at about 50℃.

(2) Place a pencil under one end of a sterile petri dish. Pour aseptically a tube of melted nutrient agar into the plate to cover the entire bottom surface, and allow it to solidify in the slanted position.

(3) After the agar has hardened, remove the pencil and place the plate flat on the table.

(4) Add 0. 1 ml of the streptomycin solution to a second tube of melted nutrient agar. Mix by rotating the tube between the palms of your hands. Aseptically pour the mixture on the surface of the gradient – plate and allow it to harden.

(5) Pipette 0. 2 ml of the *Staphylococcus aureus* culture onto the agar surface. Spread the culture over the surface of the plate using the sterile glass spreader.

(6) Incubate at 37℃ for 48h in an inverted position.

(7) Following incubation, select one or two isolated colonies present in the middle of the streptomycin concentration gradient. Streak the selected colonies toward the high – concentration end of the plate using a sterile inoculating loop.

(8) Incubate the plate at 37℃ for 48h in an inverted position.

(9) Observe the line of growth from the streaked colonies into the area of high streptomycin concentration. Growth in this area is indicative of streptomycin resistant mutants.

Results and Discussion

1. The results of drug – resistant mutation

Strains	Media	Description of growth
Staphylococcus aureus (penicillin sensitive)		
Staphylococcus aureus (penicillin resistant)		

2. The results of determination of penicillinase (I_2 – starch plate method).

Strains	Description of results
Staphylococcus aureus (penicillin sensitive)	
Staphylococcus aureus (penicillin resistant)	

3. The results of isolation of a streptomycin – resistant mutant.

4. What mechanisms are responsible for antibiotic resistance?

5. Why is it necessary to use an antibiotic gradient – plate preparation for isolation of mutants?

6. Does the streptomycin in the medium cause the mutations? Explain it.

7. Why has there been an increase in drug resistant bacterial strains in recent years?

实验三十四　营养缺陷型菌株的诱变和筛选

【实验目的】

1. 学习诱发突变和理化诱变剂的诱变作用。
2. 学习营养缺陷型的浓缩和检出方法。
3. 学习营养缺陷型的生长谱鉴定方法。

【实验原理】

突变是DNA核苷酸序列的稳定可遗传的改变。突变的发生有两条途径，自发和诱发。自发突变是在没有外来因素的作用下在细胞中以极低的频率发生。而诱发突变是微生物暴露在诱变剂中产生的。诱变剂是能够提高突变频率的物理或化学因素。

营养缺陷型是丧失合成某一物质（如氨基酸、维生素、核苷酸等）的能力而必须从环境中获得该物质的突变细菌，它们在基本培养基上不能生长，必须补充某些物质才能生长。能够在基本培养基上生长的菌株称为原养型。筛选营养缺陷型菌株主要经过如下四个步骤：诱变处理、营养缺陷型浓缩、营养缺陷型检出、生长谱鉴定。

浓缩营养缺陷型的方法有青霉素法、菌丝过滤法、差别杀菌法、饥饿法等，这些方法适用于不同的微生物。细菌的缺陷型筛选可用青霉素法，酵母菌和霉菌的缺陷型筛选可以用制霉菌素代替青霉素，放线菌和霉菌可用菌丝过滤法，它们的野生型孢子能在基本培养基中萌发并长成菌丝，缺陷型的孢子不能长成菌丝。所以把经诱变处理的孢子悬浮在基本培养液中振荡培养，在培养过程中过滤几次，每次培养时间不宜过长，这样才能充分浓缩。

营养缺陷型的检出一般有4种方法：随机筛选法（逐个点种法）、影印法、夹层培养法和限量补给法。逐个点种法，简便易行，但效率较低。

营养缺陷型的生长谱鉴定一般分两步，第一步先确定为哪一大类（氨基酸、维生素或核苷酸等）缺陷型，第二步再具体鉴定为哪种营养要求。

本实验采用化学诱变，用烷化剂硫酸二乙酯（DES）处理大肠埃希菌以获得营养缺陷型。采用青霉素法和逐个点种法进行营养缺陷型的浓缩和检出。本实验不进行大类测定，主要鉴定氨基酸缺陷型。由于氨基酸有20种，故将它们排列组合成几大组，可直接用氨基酸混合粉末点种，也可制成滤纸片贴于平板上。

【实验材料】

1. 菌种

大肠埃希菌 K12S。

2. 培养基

肉汤液体培养基，完全培养基（肉汤固体培养基），无 N 基本液体培养基，2N 基本液体培养基，50 × Vogel 盐溶液（即 50 倍浓度），平板用基本培养基。

3. 试剂

硫酸二乙酯，0.1mol/L pH7.1 磷酸缓冲液，生理盐水，20% $Na_2S_2O_3$溶液，青霉素，混合六组氨基酸（见下表）。

类别	氨基酸					
Ⅰ	赖	精	甲硫	胱	亮	异亮
Ⅱ	缬	精	苯丙	酪	色	组
Ⅲ	苏	甲硫	苯丙	谷	谷氨酰胺	脯
Ⅳ	羟脯	胱	酪	谷	天冬	天冬酰胺
Ⅴ	丙	亮	色	谷氨酰胺	天冬	甘
Ⅵ	丝	异亮	组	脯	天冬酰胺	甘

4. 器材

恒温水浴，离心机，摇床，锥形瓶，离心管，试管，吸管，吸球，平皿，牙签等。

【实验方法】

1. 诱变

（1）取一环过夜活化的斜面菌，接种于盛有 10ml 肉汤培养基的锥形瓶中，37℃摇育 5 小时。

（2）将菌液转移至离心管，3500 转/分离心 10 分钟，倾去上清液，混旋器混匀沉淀，加 pH7.1 磷酸缓冲液 10ml，放置 37℃水浴中备用。

（3）取存放于 37℃水浴中的备用菌液 2ml，加入含 8ml 磷酸缓冲液的锥形瓶中，混匀，加 0.1ml 硫酸二乙酯，迅速摇匀，立即放入 37℃，100 转/分作用 20 分钟。

（4）将处理过的菌液，加入 0.3ml 20% 硫代硫酸钠溶液以终止反应。离心集菌，菌体沉淀打匀，加入 10ml 肉汤培养基，取 1ml 用生理盐水作 10 倍系列稀释至 $10^{-4} \sim 10^{-5}$（以每只平板 50 ~ 100 个菌落为宜，故稀释度应视具体情况变动）；其余菌液于 37℃培养 2 ~ 3小时使表型表达。

（5）从 10^{-4}、10^{-5}的稀释管中，分别吸取 0.1ml，加入铺有 15ml 完全培养基的底层平板上，倾注约 5ml 融化并冷至 50℃左右的完全培养基，迅速摇匀待凝。每稀释度做 2 块平板。

（6）37℃孵箱培养 48 小时，活菌计数。平板冰箱保存，以备营养缺陷型检出用。

（7）对照处理：步骤同诱变处理，诱变剂用磷酸缓冲液代替，最后稀释至 $10^{-5} \sim 10^{-6}$，37℃培养 24 小时，活菌计数。

2. 营养缺陷型的浓缩

（1）取 5ml 表型表达过的菌液于灭菌离心管，3500 转/分离心 10 分钟。倒去上清液，沉淀打匀，加入生理盐水，离心洗涤 3 次，加生理盐水到原体积。

（2）吸取经离心洗涤的菌液 0.1ml，于 5ml 无 N 基本培养液中，37℃培养 12 小时。

（3）按1∶1加入2N基本培养液5ml，称取青霉素钠盐，使青霉素在菌液中的最终浓度约为1000 U/ml，再放入37℃培养。

（4）分别于培养12、16、24小时取样，各取0.1ml菌液倒在两个灭菌培养皿中，再分别倒入融化并冷却到50℃的基本及完全培养基，摇匀铺平待凝，放入37℃培养（培养皿上分别注明取样时间）。

3. 营养缺陷型的检出

（1）制备平板用基本培养基和完全固体培养基平板。

（2）选用经诱变、浓缩处理后长出单个菌落的若干平板，用灭菌牙签轻轻地随机取单个菌落上少量菌，先划叉点种于基本培养基，随之点种于完全培养基平板的相应位置，依次点种，每个平板可点种50个细菌。37℃培养24小时。

（3）选取在基本培养基平板上不长、完全培养基平板上生长的菌落，再在基本培养基和完全培养基上划线复证，经培养后，基本培养基上仍不生长者，可疑为营养缺陷型。

（4）将这些可疑营养缺陷型，接种于完全培养基斜面上保存，以备生长谱鉴定。

4. 营养缺陷型生长谱鉴定

（1）将可疑的营养缺陷型菌株斜面，取一环接种于装有5ml肉汤的离心管内，37℃摇育14～16小时。

（2）离心，3500转/分、10分钟，倾去上清，打匀沉淀，加生理盐水5ml，离心、洗涤2次，最后加生理盐水至原体积。

（3）吸取菌液1ml，加于已铺有固体基本培养基平板上，然后再倒入融化冷却至50℃的基本培养基约5ml，摇匀，放平待凝。也可以将菌液加入装有5ml已融化的基本培养基的试管中，迅速摇匀，即刻倾注底层平板上，铺平。

（4）在凝固的平板底上用红笔划分6区，用灭菌牙签分别挑取极少量各组混合氨基酸（每换一组必须换一根牙签），加入平板的相应位置上。

（5）37℃培养24～36小时，观察生长圈，可初步确定为哪种氨基酸缺陷型。

（6）进一步确证：将此菌按同法制成双层固体基本培养基平板，用牙签挑取该种氨基酸粉末少许，点于平板中间，经培养后，在该氨基酸周围出现明显生长圈。

【结果与讨论】

1. 诱变作用结果记录

试验组	稀释度	加菌液量（ml/皿）	平均菌落数（个/皿）	菌数/ml	死亡率（%）
对照	10^{-5}				
	10^{-6}				
诱变处理	10^{-4}				
	10^{-5}				

2. 生长谱鉴定结果记录

菌株编号	生长组别	营养要求

3. 用化学诱变剂处理细菌，为什么要用缓冲液来制备菌悬液？

4. 为什么诱变处理要在充分振荡的条件下进行？

5. 营养缺陷型的浓缩有哪些常用方法？

6. 用逐个点种法筛选营养缺陷型时，要先点种在基本培养基平板上，然后再点种在完全培养基平板上，为什么？

34 Mutagenesis and Isolation of Auxotroph

Purpose

1. To learn how to induce mutations through mutagenesis with mutagen.
2. To learn how to detect and isolate auxotroph.
3. To learn how to identify auxanograms of auxotroph.

Principle

A mutation is the result of a stable, heritable change in the nucleotide sequence of DNA. Mutations may occur in one of two ways, spontaneous or induced. Spontaneous mutations arise in a very low rate in all cells and develop in the absence of any external agent. Induced mutations, on the other hand, are results of exposure of the microorganism to mutagen. Mutagen is physical or chemical agent that can raise mutation rate.

An auxotroph is a mutated bacterium that lacks the ability to synthesize an essential nutrient such as amino acids, vitamins and nucleotides and must obtain it from its surroundings. That is, the mutant can grow on medium containing the nutrient supply, but not on minimal medium. Microbial strains that can grow on minimal medium are called prototrophs.

There are four main procedures to acquire an auxotroph: mutagenesis, enrichment, detection and isolation, auxanograms identification.

The methods to eliminate prototroph and enrich auxotroph include penicillin enrichment method, filtration enrichment method, selective bactericidal method and starvation method and so on. They can be used to enrich different microorganisms. To enrich auxotroph of bacteria penicillin method can be used. Mycostatin is used instead of penicillin in the case of yeast and mold. Filtration method is used to enrich auxotroph of actinomycetes and mold. The prototroph spores of actinomycetes and mold can germinate to hypha while the auxotroph spores not. Incubate the spores after mutagenesis in minimal broth and filtrate the culture several times every short period. So the auxotroph will be enriched fully.

The usual methods of detection and isolation of auxotroph are random isolation (one by one isolation), replica plating, layer plating and limited supply method. The random method is easy to do but its efficiency is low.

The auxanogram identification of auxotroph includes two steps. The first is to define which nutrient (amino acids, vitamins or nucleotides) the auxotroph cannot synthesize. The second is to define which specific amino acid or others it needs to be added.

In this experiment, DES, one kind of alkylating agent, will be used to induce mutation in *Escherichia coli* to acquire auxotroph. Penicillin enrichment and random isolation method will be used to enrich and isolate auxotroph. In auxanogram identification just amino acid auxotroph will be defined. The twenty kinds of amino acids are arranged and composed to six groups and drop the mixture powder directly or place the mixture filter paper on the plate.

Materials and Apparatus

1. Strains

E. coli K12S.

2. Media

Nutrient broth, nutrient broth agar (complete medium, CM), minimal broth containing no N source, minimal broth containing two - fold N source, 50 × Vogel salt solution (*i. e.* enriched 50 - fold), minimal agar (minimal medium, MM) used in plate.

3. Reagents

Diethyl sulfate (DES), 0. 1mol/L pH7. 1 phosphate buffer (PBS), normal saline (NS), 20% $Na_2S_2O_3$ solution, penicillin, mixture of amino acids (see the table below).

Group	Amino acids					
Ⅰ	Lys	Arg	Met	Cystine	Leu	Ile
Ⅱ	Val	Arg	Phe	Lys	Trp	His
Ⅲ	Thr	Met	Phe	Glu	Gln	Pro
Ⅳ	Hyp	Cystine	Tyr	Glu	Asp	Asn
Ⅴ	Ala	Leu	Trp	Gln	Asp	gly
Ⅵ	Ser	Ile	His	Pro	Asn	Gly

4. Apparatus

Water bath, centrifuge, rotator, flask, centrifuge tubes, tubes, pipettes, pipetter, petri dishes, toothpicks.

Procedures

1. Mutagenesis

(1) Transfer a loopful of overnight agar slant culture and inoculate it to a flask containing 10ml broth. Incubate the flask with shaking at 37℃ for 5h.

(2) Transfer the suspension to a centrifuge tube. Centrifuge at 3500 rpm for 10 min. Discard the supernatant and resuspend the cells in 10ml of PBS. Store in 37℃ water bath.

(3) Transfer 2ml of the suspension in 37℃ water bath to a flask containing 8ml of PBS and mix well. Add 0. 1ml DES to the flask and mix quickly by rotating and incubate in 37℃ water bath with shaking at 100 rpm for 20 min.

(4) Add 0. 3ml $Na_2S_2O_3$ solution (20%) into the flask to stop the mutagenesis. Centrifuge and harvest the cells. Resuspend the cells in 10 ml nutrient broth. Transfer 1 ml and dilute serially at 10 – fold with NS to 10^{-4} and 10^{-5} (The dilution degree may be changed to make 50 to 100 colonies grow on every plate) . Incubate the rest culture at 37℃ for 2 to 3h to let the phenotype express.

(5) Transfer 0. 1ml from the 10^{-4} and 10^{-5} tubes to the plates containing 15ml nutrient agar bottom. Pour 5 ml of nutrient agar that has been already melted and cooled to about 50℃ onto the bottom plate and mix quickly. Let it solidify. Make 2 plates every dilution degree.

(6) Incubate all the plates at 37℃ for 48h. Count the colonies.

(7) Add 0. 1ml PBS instead of DES to the control flask. The dilution degrees are 10^{-5} and 10^{-6}. Incubate at 37℃ for 24h and count the colonies. The other steps are same as to the mutagenesis case.

2. Enrichment of Auxotroph

(1) Transfer 5ml culture after phenotype expression to a centrifuge tube. Centrifuge at 3500rpm for 10min.

(2) Discard the supernatant. Wash and centrifuge with NS for 3 times. Then resuspend the cells to the original volume.

(3) Transfer 0. 1ml of suspension to 5ml of minimal broth containing no N source. Incubate at 37℃ for 12h.

(4) Add 5ml minimal broth containing 2 – fold N source to the culture (*V/V* = 1 : 1) . Add penicillin sodium to the culture to make the concentration be about 1000 U/ml. Incubate at 37℃.

(5) Take samples at the 12th, 16th, 24th hour. Transfer 0. 1ml to 2 petri dishes and pour MM and CM melted and cooled to about 50℃, respectively. Mix well rapidly and let them harden. Mark the sample time on the plates and incubate at 37℃.

3. Detection and Isolation of Auxotroph

(1) Prepare MM and CM plates.

(2) Select the plates on which the colonies are single after mutagenesis and enrichment. Transfer randomly the single colony by stabbing the colony with the sterile toothpick. Streak a cross on the surface of MM plate and then on the same area of CM

plate. Inoculate in turn. There are about 50 inoculation crosses on every plate. Incubate at 37℃ for 24h.

(3) Select the colony grows on the CM plate but not grow on the MM plate. Streak again on the MM and CM plates. After incubation the colony grows on CM and still not grow on MM may be auxotroph.

(4) Inoculate the suspicious auxotroph to CM agar slant and store for auxanogram identification.

4. Auxanogram Identification of Auxotroph

(1) Transfer a loopful the suspicious auxotroph CM agar slant culture to a centrifuge tube containing 5 ml broth and incubate with shaking at 37℃ for 14 to 16h.

(2) Centrifuge at 3500 rpm for 10 min. Discard the supernatant. Wash and centrifuge for 2 times with NS. Then add NS to 5 ml.

(3) Transfer 1 ml of the suspension to the bottom of MM plate and pour about 5 ml of MM melted and cooled to about 50℃. Mix well rapidly and let harden. The suspension can also be transferred to a tube containing 5 ml of MM melted and cooled to about 50℃. Mix well rapidly and pour it down onto the plate and let it harden.

(4) Divide the plate into 6 areas on the bottom surface using a red wax pencil. Pick the mixture powder of amino acids of the 6 groups using toothpick, on the corresponding area respectively (one toothpick for each group only).

(5) Incubate at 37℃ for 24 to 36h. Examine the growth circles and identify - which amino acid the auxotroph cannot synthesize.

(6) To identify further: Make the double – layer plate containing the auxotroph by the above method. Pick the amino acid powder onto the center of the plate. After incubation, the growth circle surrounding the amino acid will be obvious.

Results and Discussion

1. The results of mutagenesis.

Group	Dilution degree	Add culture (ml/dish)	Average number of colonies/plate	Number of bacteria/ml	Mortality (%)
Control	10^{-5}				
	10^{-6}				
Mutagenesis	10^{-4}				
	10^{-5}				

2. The results of auxanogram identification.

Strain No.	Group No. of Amino Acid	Amino Acid requirement

3. When treated the bacteria with chemical mutagen, PBS should be used to prepare the bacterial suspension, why?

4. Why should the mutagenesis be done under the shaking condition?

5. Which method are there to enrich the auxotroph?

6. It should be inoculate first on the MM plate and then on the CM plate when isolating auxotroph using random isolation method, why?

实验三十五　Ames 试验检测化学诱变剂

【实验目的】

学习 Ames 试验的基本原理和方法。

【实验原理】

Ames 等人用鼠伤寒沙门菌的组氨酸营养缺陷型突变株，观察其在测试物作用下，回复突变为原养型而建立了化学诱变剂的检测系统。所用的菌株是几种不同的组氨酸突变型的标准测试菌株，它们除了组氨酸突变（his^-）外，还有 DNA 切除修复系统的缺失突变（$\Delta uvrB$）和细菌表面脂多糖屏障突变（*rfa*）。

由于化学物质对微生物的诱变作用，反映了它对哺乳动物的潜在的致癌作用，所以可用此系统来检测一些化学物质的诱变性。该法简便、快速、有一定可靠性，是检测农药、药物、食品添加剂以及工厂、实验室和环境中的一些化学物质诱变性的好方法。

【实验材料】

1. 菌种

TA97 移码突变（*rfa* ΔuvrB *his*D6610，R 因子：抗氨苄青霉素 Ap^r）

TA98 移码突变（*rfa* ΔuvrB *his*D3052，R 因子：抗氨苄青霉素 AP^r）

TA100 碱基置换（*rfa* ΔuvrB *his*G46，R 因子：抗氨苄青霉素 AP^r）

TA102 赭石型突变（*rfa* Ap^r *his*G428，R 因子：抗 Ap^r 和抗四环素 Tc^r）

2. 培养基

50 × Vogel 盐溶液（即 50 倍浓度），平板用基本培养基，上层半固体培养基，肉汤液体培养基，固体完全培养基。

3. 试剂

（1）药物：氨苄青霉素溶液（8mg/ml），四环素溶液（8mg/ml），结晶紫溶液（0.1%），正定霉素溶液，亚硝基胍溶液，丝裂霉素 C 溶液，2－氨基芴溶液。

（2）大鼠肝匀浆（S－9 部分），NADP（辅酶Ⅱ）和 G－6－P（葡萄糖－6－磷酸钠盐）使用液。

4. 器材

水浴，15W 紫外灯，摇床，培养箱，镊子，培养皿，锥形瓶，吸管，试管等。

【实验方法】

（一）菌株遗传标记的鉴定

1. 组氨酸需要

（1）制备含生物素 3μmol/L 和同时含生物素 3μmol/L 及组氨酸 260μmol/L 的平板用基本培养基平板各一个。

（2）将过夜培养的各菌株培养物分别取一环先于含生物素平板上划一道，然后在含生物素和组氨酸平板上划一道。

（3）平板于37℃培养过夜，检查2种平板上各株菌的生长情况。

2. *rfa* 突变

（1）测试菌过夜培养物0.1ml，加入装有3ml，恒温于45℃的半固体培养基试管内，摇匀，迅速倾倒于铺有固体完全培养基底层平板上，使上层琼脂均匀分布，待凝。

（2）用灭菌镊子取无菌滤纸片，在结晶紫溶液中浸湿后，放置在上述平板中央，37℃培养过夜，观察滤纸片周围有无出现抑菌圈。

3. Δ*uvrB* 突变

（1）制备固体完全培养基平板一个。

（2）分别取各测试菌液一环，平行划线于平板表面。平板一半用黑纸遮盖，另一半置于15W紫外灯下，距离33cm，照射8秒。避光，放置37℃培养24小时。具有Δ*uvrB*缺陷的菌株，对紫外线敏感，故只有未照射的一边平板上生长。

4. R因子菌株（抗氨苄青霉素试验）

（1）制备固体完全培养基平板一个。

（2）取一环氨苄青霉素溶液在平板上涂一条带。

（3）待药液干后，将各测试菌液分别垂直于药液带接种一条，待干后，37℃培养24小时，观察在氨苄青霉素带的周围是否有生长抑制区。

（二）诱变剂测试

1. 点试法

（1）灭菌培养皿，倒入融化并冷至50℃左右的平板用基本培养基约15ml作为底层，铺平待凝。

（2）吸取0.1ml测试菌的新鲜过夜培养物于装有3ml上层半固体培养基试管中，摇匀，迅速倾注底层平板上，铺平待凝。

（3）分别将蘸有化学诱变剂NTG（1mg/ml）、正定霉素（20mg/ml）和丝裂霉素C（2.5mg/ml）的滤纸片，放入各平板中央。37℃培养48小时，观察结果。

2. 平板掺入法

（1）制备底层平板用基本培养基平板2个。

（2）装有2ml上层半固体培养基试管内，加0.1ml菌液，摇匀，倾注于底层平板上作对照。

（3）吸取0.1ml菌液和0.1ml待检物于同一小试管内，混匀（也可放置37℃，振荡作用20分钟，使充分接触），随后加入2ml上层半固体培养基，摇匀，迅速铺平于底层平板上。37℃培养48小时，计数。如回复突变数超过阴性对照（自发回复突变数）2倍以上，即为阳性。各菌株自发回复突变菌落数正常范围如下表。

菌株	TA97	TA98	TA100	TA102
自发回变菌落数	90～180	30～50	120～200	240～320

基本方法同平板掺入法，只是在上层培养基中还需要加入S－9混合液0.2ml，该混合液必须现用现配，把低温贮存的S－9置室温下融化，每2ml加入10ml的NADP和G－6－P使用液。混合液置冰浴中，用后多余部分弃去，下次实验时重配。

【结果与讨论】

1. 菌株遗传标记鉴定（描述现象）。

鉴定 菌株	组氨酸	结晶紫	紫外线	R因子
TA97				
TA98				
TA100				
TA102				

2. 点试法

以图表示各部位的含义，写出菌株及诱变剂。

3. 平板掺入法

测试物名称	浓度（mg/ml）	测试菌	测试物菌落数/皿	自发回变菌落数/皿
正定霉素	2	TA97		
正定霉素	2	TA98		
NTG	0.1	TA100		
丝裂霉素	0.005	TA102		
2－氨基芴 （加S－9混合液检测）	0.4	TA97		
		TA98		
		TA100		
		TA102		

结论：

4. Ames试验中加入S－9的目的是什么？

5. Ames试验中组氨酸－生物素溶液的作用是什么？

6. 用细菌系统替代哺乳动物系统检测化学致癌剂的优点和缺点各是什么？

7. 滤纸片周围可观察到抑菌圈。你能解释原因吗？如果出现这种情况，你怎样判断该化合物是否致突变？

35 Ames Test: A Bacterial Test System for Detecting Mutagens and Carcinogens

Purpose

To learn how to screen potential chemical carcinogens using a bacterial test system.

Principle

Various chemical compounds such as industrial pollutants, pesticides, food additives, hair dyes, cigarette smoke, increase the rate of mutations. It is important to identify such mutagens because they present a potential health risk. Additionally, the carcinogenic (cancer-causing) potential of many compounds to which we are exposed in our environment is highly correlated with their ability to induce mutation. From a genetic aspect there is strong evidence linking carcinogenicity to mutagenicity. Research indicates that approximately 90% of the chemicals proved to be carcinogens are mutagens. They cause cancer by inducing mutations in somatic cells.

Because of the ease with which large bacterial populations can be handled, the mutagenic potential of various chemicals can be demonstrated most easily by using bacterial cell systems. Despite the fact that mammalian cell structure and human enzymatic pathways differ from those in bacteria, the chemical nature of DNA is common to all organisms. This permits the use of bacterial test systems for the rapid detection of possible mutagens and therefore possible carcinogens.

The Ames test, developed by Bruce Ames in the 1970s, which has been widespread used, is a simple and inexpensive procedure that uses a bacterial test organism to screen for mutagens. The Ames test is a mutational reversion assay employing several special strains of *Salmonella typhimurium*, each of which has a different mutation in the histidine biosynthesis operon. The test organism is a histidine-negative (his^-) and biotin-negative (bio^-) auxotroph that will not grow on a medium deficient in histidine unless a back mutation to his^+ (histidine-positive) has occurred. The bacteria also have mutational alterations of their cell walls (*rfa*) that make them more permeable to test substances. To further increase assay sensitivity, the strains are defective in the ability to carry out excision repair of DNA and have plasmid genes that enhance error-prone DNA repair ($\Delta uvrB$).

The Ames test generally requires the addition of a liver homogenate, S－9, to make this bacterial system more comparable to a mammalian test system. Because many compounds must be activated before they demonstrate carcinogenicity, the reliability of the Ames *Salmonella* test is enhanced by the use of a liver enzyme system which may convert inactive substances to carcinogenic forms as sometimes occurs in animal metabolism.

In the Ames test the tester strains are plated with the substance being tested and the appearance of visible colonies followed. To ensure that DNA replication can take place in the presence of the potential mutagen, the bacteria and test substance are mixed in dilute molten top agar to which a trace of histidine has been added. This molten mix is then poured on top of minimal agar plates and incubated for 2 to 3 days at 37℃. All of the histidine auxotrophs will grow for the first few hours in the presence of the test compound until the histidine is depleted. Once the histidine supply is exhausted, only revertants that have mutationally regained the ability to synthesize histidine will grow. The visible colonies need only be counted and compared to controls in order to estimate the relative mutagenicity of the compound; the more colonies, the greater the mutagenicity.

Materials and Appatatus

1. Strain

Salmonella typhimurium, TA97, TA98, TA100, TA102.

2. Media

Minimal medium, top agar medium, nutrient agar, nutrient broth.

3. Reagent

Sterile biotin－histidine solution, ampicillin solution (8mg/ml), tetracycline solution (8mg/ml), crystal violet solution (0.1%), daunomycin solution, NTG solution, 2－nitrofluorene solution, MMC solution, S－9 fraction, NADP and G－6－P solution.

4. Apparatus

UV lamp, water bath, sterile pipettes, tubes.

Procedures

A. Identifying the Genetic Marker of Test Strains

1. Histidine－ require test

(1) Prepare minimal agar plates containing biotin (3μmol/L) and minimal agar plates containing biotin (3μmol/L) and histidine (260μmol/L).

(2) Using sterile inoculating loop, transfer a loopful of overnight broth culture

and streak out on the surface of the plates to form a line. Streak the plate of biotin first, then the plate containing both of biotin and histidine.

(3) Incubate at 37℃ for 48h in an inverted position.

(4) After incubation, examine the growth on the two kinds of plates.

2. *rfa* mutation test

(1) Prepare nutrient agar plate.

(2) Melt tubes of 3 ml top agar in a hot – water bath and maintain the molten agar at 45℃.

(3) Add 0. 1ml overnight broth culture to the tube and mix quickly by rotating it between your palms.

(4) Pour the top agar cultures onto the nutrient agar plates and allow them to solidify.

(5) Using sterile forceps, dip filter paper discs into crystal violet and place them in the center of the plates, respectively.

(6) Incubate at 37℃ overnight.

(7) Examine the inhibition zone surrounding the disc.

3. Δ*uvrB* mutation test

(1) Prepare nutrient agar plate.

(2) Transfer a loopful of culture and streak out on the surface of nutrient agar plate to form a line. Shelter ½ plate with black paper and expose another ½ plate under the UV lamp at a distance of 33 cm for 8 s.

(3) Incubate at 37℃ for 24h in an inverted position, keeping from light.

(4) After incubation, examine the growth on the plate.

4. R – plasmid

(1) Prepare nutrient agar plate.

(2) Transfer a loopful of ampicillin solution and draw a band on the plate and allow it dry.

(3) Inoculating the cultures vertical to the drug band to form lines.

(4) Incubate at 37℃ for 24h in an inverted position.

(5) Observe the growth and examine the inhibition zone surrounding the drug band.

B. Mutagen Test

1. Spot method

(1) Prepare minimal agar plates.

(2) Melt tubes of 3 ml top agar in a hot – water bath and maintain the molten agar at 45℃.

(3) Add 0. 1ml overnight broth culture to the tube and mix quickly by rotating it between your palms.

(4) Pour the mixture onto the nutrient agar plate and allow it to solidify.

(5) Using sterile forceps, dip filter paper discs into the chemical mutagens and place them in the center of the plates, respectively.

(6) Incubate at 37℃ for 48h.

(7) Observe the area around each disc for revertant colonies. Scattered colonies of revertant mutants will appear over the surface of the plate. A positive result is indicated by a relatively high concentration of colonies surrounding the disc.

2. Plate method

(1) Prepare minimal agar plates.

(2) Melt tubes of 2 ml top agar in a hot – water bath and maintain the molten agar at 45℃.

(3) Add 0. 1ml overnight broth culture to the tube and mix quickly by rotating it between your palms. Pour the mixture onto the nutrient agar plate and allow it to solidify. This plate is negative control. The spontaneous revertant ranges are: TA97: 90 ~ 180, TA98: 30 ~ 50, TA100: 120 ~ 200, TA102: 240 ~ 320.

(4) Transfer 0. 1ml of culture and 0. 1 ml chemical mutagens to a tube. Mix and place on the rotator at 37℃ for 20 min.

(5) Add the mixture to the top agar tube and mix quickly by rotating it between your palms. Pour the mixture onto the nutrient agar plate and allow it to solidify.

(6) Incubate at 37℃ for 48h. After incubation, count the colonies of revertant.

3. Adding S – 9 method

The method is almost same as plate method above. The only difference is to add 0. 2 ml S – 9 mixture to top agar tube. S – 9 mixture must be freshly prepared.

Results and Discussion

1. The results of identifying the genetic marker of test strains.

Identification / Strain	Histidine	*rfa* mutation	*ΔuvrB*	R – plasmid
TA97				
TA98				
TA100				
TA102				

2. The results of spot test: illustrate each site of the plate and record the strain and mutagen.

3. The results of plate method.

Test chemical	Concentration (mg/ml)	Strain	Number of induced mutations/plate	Number of spontaneous mutations/plate
Daunomycin	2	TA97		
Daunomycin	2	TA98		
NTG	0.1	TA100		
MMC	0.005	TA102		
2 - AF (Add S - 9 mixture)	0.4	TA97		
		TA98		
		TA100		
		TA102		

4. What is the purpose of adding the S - 9 in the Ames test?

5. What is the purpose of the biotin - histidine solution in the Ames test?

6. What are the advantages and disadvantages of applying bacterial systems instead of mammalian systems to test for chemical carcinogenicity?

7. An inhibition zone may be observed surrounding the filter paper disc. Can you explain why might this happen? If it happened, what would you do to determine whether the compound was mutagenic?

实验三十六 抗药质粒接合传递实验

【实验目的】

学习抗药质粒接合传递的基本原理和方法。

【实验原理】

供体菌痢疾杆菌 D15，对四环素、链霉素和氯霉素抗药，是多剂抗药菌，由抗药质粒编码。受体菌大肠埃希菌 K12W1485F$^-$，对利福平抗药，由染色体编码。供、受体菌在一定条件下混合，随之接合，接合子在含选择性标记氯霉素（或链霉素或四环素）和利福平的 EMB 培养基上生长，形成具有紫黑色金属光泽的菌落，证明供体菌抗药质粒转移至受体菌。而二亲本在上述含药平板上均不生长。接合子可进一步通过药敏试验和电泳检测抗药质粒的存在来证明。

【实验材料】

1. 菌种

（1）供体菌　痢疾杆菌 D15（R 因子，为氯霉素、四环素、链霉素抗药）。

（2）受体菌　大肠埃希菌 K12W1485（利福平抗药）。

2. 培养基

肉汤培养基，EMB 培养基。

3. 试剂

氯霉素（3mg/ml），利福平（2mg/ml）。

4. 器材

无菌吸管，小试管，平皿等。

【实验方法】

1. 活化

将二亲本分别接种肉汤培养基斜面，37℃培养过夜。然后分别取少量菌转种于盛有 1ml 肉汤试管内，37℃培养 5 ~6 小时。

2. 接合传递

（1）用吸管吸取 1ml 肉汤，置于无菌小试管内，同时加入供、受体菌培养物各 0. 1ml，混匀。37℃培养 2 小时使之接合。

（2）制备 EMB 含药平板：吸取氯霉素溶液 0. 15ml 于平皿中，倾注 15ml 融化并冷至 50℃左右的 EMB 培养基，迅速摇匀，待凝，氯霉素作用浓度为 30μg/ml。吸取利福平溶液 0. 23ml 于平皿中，倾注 15ml 融化并冷至 50℃左右的 EMB 培养基，迅速摇匀，待凝，利福平作用浓度为50μg/ml。吸取氯霉素溶液0.15ml和利福平溶液0.23ml于同一平皿中，倾注15ml融化并冷至50℃左右的EMB培养基，迅速摇匀，待凝。

（3）将EMB含药平板背面，用红笔划分三个区域，吸取0.1ml混合菌液涂布于其中一大区域，另二区域分别用接种环取一环接种供、受体菌，作为对照，37℃培养过夜。

（4）在不含药的EMB平板上划分三个区域，分别接种二亲本和混合菌，培养过夜。

（5）观察供、受体菌和接合子的生长情况，注意观察菌落色泽。

【结果与讨论】

1. 观察平板上细菌的生长状况，特别是菌落的色泽，记录实验结果。

菌株 / EMB	D15	W1485	接合子
含利福平平板			
含氯霉素平板			
含氯霉素及利福平平板			
无药平板			

2. 怎样判定抗药基因是质粒编码还是染色体编码？

3. 本实验中怎样选择接合子？

4. 实验中形成了多少双重抗药菌的菌落？发生了多少质粒转移？质粒转移是否会随着培养时间的延长而增加，为什么？

5. 本实验中利福平抗性会从大肠埃希菌转移到痢疾杆菌吗？为什么？

36 Bacterial Conjugation: The Transfer of Antibiotic – Resistant Plasmids

Purpose

To be familiar with the principles and methods of the transfer of antibiotic – resistant plasmids through conjugation.

Principle

Plasmids that can affect the conjugation between individual bacteria are termed conjugative plasmids. These plasmids carry genes that code for the ability to transfer themselves from the donor host bacterium to a suitable recipient bacterium by conjugation. The donor strain is D15, which is a multiple drug resistant strain that is resistant to chloramphenicol (CM), tetracycline and streptomycin. Its resistance determinant is located on a conjugative plasmid. The recipient strain is *E. coli* K12 W1485 that carries a chromosomal gene for rifampin (Rif) resistance. Mix the two strains and let conjugation occur. On the EMB plate containing both chloramphenicol (streptomycin or tetracycline) and rifampin the conjugant will form metallic sheen purple – black colonies. Neither the donor nor the recipient can grow on this plate. The conjugant can be detected by drug – sensitivity test and its R plasmid can be detected by electrophoresis.

Materials and Apparatus

1. Strains

Donor strain: Shigella dysenteriae D15 (R plasmid: resistant to chloramphenicol, tetracycline and streptomycin) . recipient strain: *E. coli* K12 W1485 (resistant to rifampin) .

2. Media

Nutrient broth, EMB agar.

3. Drug

Chloramphenicol (CM) solution (3mg/ml), rifampin (Rif) solution (2mg/ml).

4. Apparatus

Petri dishes, tubes, pipette, glass spreader.

Procedures

1. Activation

Inoculate the donor and recipient strains into nutrient agar slant and incubate at 37℃ overnight, respectively. Then inoculate them to 1ml broth tubes and incubate at 37℃ for 5 ~ 6h, respectively.

2. Prepare the EMB plates

(1) 1EMB agar plate not containing drug pour 15ml of EMB agar that has been already cooled to about 50℃ into the petri dish bottom and mix quickly. Let agar solidify.

(2) 1EMB agar plate containing 30 μg CM per ml Transfer 0. 15 ml of CM solution and pour 15ml EMB agar.

(3) 1EMB agar plate containing 50 μg Rif per ml Transfer 0. 23 ml of Rif solution and pour 15ml EMB agar.

(4) 1EMB agar plate containing CM (30 μg/ml) and Rif (50 μg/ml) Transfer 0. 15ml of CM solution and 0. 23ml of Rif solution and pour 15ml EMB agar.

Divide the plates into three areas by drawing lines on the bottom surface of the plates.

3. Conjugation

(1) Prepare the mating mixture by aseptically transferring 0. 1ml of a broth culture of donor and recipient strains respectively into 1ml of sterile broth. Mix and incubate at 37℃ for 2h.

(2) Transfer a drop of donor culture to one area and a drop of recipient culture to another area of the plates.

(3) Remove the mating mixture from the incubator and transfer 0. 1ml to the third area of the plates. Carefully spread it over the area using the glass spreader.

(4) Incubate invertedly at 37℃ overnight.

Results and Discussion

1. Examine all plates for growth especially the color of the colonies and record the results in the report.

Plate	D15	W1485	Conjugant
CM			
Rif			
CM + Rif			
No drug			

2. How to distinguish the bacterial drug – resistance encoded by plasmid or by chromosomal gene?

3. How to select conjugant in this exercise?

4. How many doubly resistant colonies were formed? About how many plasmid transfers took place? Should this number increase with incubation time? Why or Why not?

5. Is it likely that rifampin resistance will be frequently transferred form *E. coli* W1485 to D15? Why?

实验三十七　抗药质粒 DNA 的分离及检测和细菌转化

【实验目的】

1. 学习质粒 DNA 分离的常用方法和基本原理。
2. 学习琼脂糖凝胶电泳技术检测质粒 DNA 的基本方法。
3. 学习用质粒转化细菌的基本方法和原理。

【实验原理】

质粒是染色体外的遗传因子，是共价闭合环状的双链 DNA 分子，能够在细胞质中独立自主地进行复制，并稳定地遗传给子代细胞，赋予细胞以某些性质和生物功能。本实验用的大肠埃希菌携带 pBR322 抗药质粒，使该菌对四环素和氨苄青霉素抗药。根据质粒的抗药性较容易检出和鉴定。pBR322 是遗传工程中常用的基因载体。

分离质粒 DNA 的方法很多，都是利用质粒与染色体 DNA 在特性和构型上的不同而进行的，包含三个基本步骤：①培养菌体和扩增质粒。②收获和裂解菌体。③纯化质粒 DNA。本实验采用煮沸分离法，利用溶菌酶和表面活性剂等的作用，破坏细胞壁中的糖肽层，同时使细胞膜崩解，从而达到菌体的充分裂解，此时细菌染色体可以形成卷曲状折叠结构，附着在细胞膜碎片上，随后通过沸水浴快速作用，使蛋白质迅速凝固并将染色体包裹，经离心呈胶状沉淀物而去除，质粒 DNA 存在于上清液中，加入冷无水乙醇，在 -20℃放置 30 分钟，可使 DNA 析出。所获质粒 DNA 可用于转化及酶切等。

质粒 DNA 的检测一般采用琼脂糖凝胶电泳。DNA 分子在高于其等电点的 pH 溶液中带负电荷，在电场中向正极移动。DNA 分子在电场中，通过凝胶介质而泳动，除电荷效应外，凝胶介质还有分子筛效应。泳动速度与分子大小及构象有关。分子量相同而构型不同的质粒 DNA 电泳迁移位置明显不同。为了检测 DNA，一般用溴乙锭（EB），其原理是 EB 可插入 DNA 双螺旋结构的二个碱基之间，它和核酸形成一种荧光络合物，在 254 nm 波长紫外光照射下，呈现橙红色的荧光。

转化是将供体 DNA 分子，引入受体细胞，以传递遗传信息的一种手段。目前转化常用于微生物遗传学，分子遗传学及基因工程等方面。转化时受体细胞需处于感受态，一般用氯化钙处理处于对数早期受体菌，使获得感受态。此时细胞膜的通透性发生变化，DNA 分子易进入细胞，实现了遗传信息转移。这种被供体 DNA 所转化的受体细胞叫做转化子。本实验用 pBR322 质粒 DNA 转化大肠埃希菌 K12 802，使获得的转化子对四环素和氨苄青霉素抗药。

【实验材料】

1. 菌种

大肠埃希菌 C600（pBR322），大肠埃希菌 K12 802。

2. 培养基

LB 培养基，M9 培养液，固体完全培养基。

3. 试剂

STET 液，溶菌酶液，TE 液，电泳液，0.7% 琼脂糖，溴酚蓝－甘油溶液，EB 溶液（10μg/ml），60mmol/L $CaCl_2$，4mg/ml 氨苄青霉素溶液，5mg/ml 四环素溶液，生理盐水，无水乙醇。

4. 器材

无菌牙签，离心管，台式高速离心机，电泳仪，电泳槽，毛细管，254nm 紫外分析仪，恒温水浴，摇床，离心机，吸管，试管，平皿，涂布棒等。

【实验方法】

1. 质粒 DNA 的分离

（1）取新鲜培养过夜的活化斜面菌约 1mm^3 菌量，在含 100μl 的 STET 液的小离心管内磨匀。加溶菌酶液 8μl，室温放置 5 分钟。放入 100℃的沸水浴中，作用 60 秒。随即迅速放入 0℃的冰水浴中 5 分钟。

（2）8000 转/分离心 5 分钟。用无菌牙签挑出胶状沉淀物，弃去。

（3）加入预冷的无水乙醇约 100μl，混匀，－20℃放置 30 分钟，10000 转/分离心 5 分钟。

（4）倾去乙醇，即得 DNA，加入 20μl TE 液使 DNA 溶解，即可用电泳检测。

2. 质粒 DNA 的检测

（1）琼脂糖凝胶制备　称取 0.7g 琼脂糖置于锥形瓶中，加入电泳液 100ml，盖上棉塞，121℃10 分钟，融化备用。

（2）凝胶板制备　先将融化的琼脂糖凝胶用毛细管吸取少量滴加在玻璃板四周，封边。再将冷却至 50℃的琼脂糖凝胶迅速倒入，使成厚约 3mm 的凝胶层。室温放置 1 小时，待胶液全部凝结后，轻轻拔出试样及挡板。电泳槽内倒入电泳液。

（3）加样　吸取 40μl pBR322 质粒 DNA，与 10μl 溴酚蓝－甘油液混匀，加入样品槽内，该染料在电泳过程中作为前沿指示剂。

（4）电泳　接通电源，使电压恒定为 80V，使溴酚蓝液向正极移动。

（5）注意观察，指示剂前沿走至所需距离时，停止电泳。取出凝胶板于 254nm 的紫外灯下观察。

（6）画出点样孔，染色体，质粒带，RNA 及溴酚蓝位置（本法染色体去除较彻底，可能看不见染色体带）。

3. 细菌转化

（1）取一环斜面菌，接种于含 5ml M9 培养液的试管内，37℃振荡培养过夜。

（2）吸取 0.5ml 菌液，加入含 5ml 新鲜 M9 培养液的锥形瓶中，37℃摇育 2 小时，离心集菌，3500 转/分 15 分钟。用预冷的灭菌生理盐水洗涤一次，沉淀物加 0.5ml 预冷的 60 mmol/L $CaCl_2$溶液混匀，即得感受态菌液。

（3）吸取感受态菌液 0.1ml 加入 0.1ml 已稀释的 DNA（根据预试验结果），冰浴放置 30 分钟，随即放入 42℃水浴约 4 分钟，再冰浴 1 小时。

（4）加入 2 倍体积（约 0.4ml）M9 培养液，37℃表型表达 3 小时。

（5）适当稀释，取某稀释度 0.1ml 涂布于含氨苄青霉素（终浓度 20μg/ml）和四环素（终浓度 25μg/ml）的完全固体培养基上。同时取不加 DNA 的菌液及 DNA，在相应平板上作对照实验，37℃培养过夜。

【结果与讨论】

1. 转化实验结果

培养基 / 实验组	完全培养基上生长情况（含氨苄青霉素和四环素）
受体菌	
质粒 DNA	
转化组	

2. 怎样制备感受态的大肠埃希菌？
3. 本实验中若省略表型表达结果会有不同吗？为什么？
4. 在进行正式转化实验前需测定受体菌的生长曲线，为什么？

37 Isolation and Characterization of Plasmid DNA and Bacterial Bacterial Transformation

Purpose

1. To be familiar with isolating plasmid DNA.
2. To learn how to characterize the size of DNA you isolate by gel electrophoresis.
3. To learn how to form recombinant bacterial cells by transformation.

Principle

Some microorganisms contain plasmids, which are relatively small extra - chromosomal genetic elements. Plasmids are covalently closed and circular (CCC) DNA molecules that permit the cell to store additional genetic information beyond that is encoded on the chromosome. In general, plasmids do not contain genetic information for the essential metabolic activities of the microorganism but rather for specialized features. Plasmids can replicate themselves and many can be exchanged between cells, permitting the transfer of genetic information from one microbial population to another. The fact that plasmids are self - reproducing, relatively small, and can be transferred from one cell to another makes them useful tools in genetic engineering, which has become the technology for constructing microorganisms with new capabilities.

E. coli C600 used in this exercise carries pBR322 which is antibiotic - resistant plasmid. So the bacterium is resistant to tetracycline and ampicillin and can be isolated and detected. pBR322 is the common carrier in genetic engineering.

In genetic engineering, plasmids carry unrelated DNA into recipient cells, which then become "engineered" cells. After being isolated from their host cells, the circular DNA of the plasmid is cut open (nicked) with enzymes, the restriction endonucleases, creating sites where foreign donor DNA can be inserted. Once the new DNA has been inserted into the plasmid DNA, other enzymes, the ligases, seal the nicked ends of the DNA to recreate a circular form. In this exercise, you will follow some of the procedures used to clone genes in genetic engineering technology. First, plasmid DNA from *E. coli* cells will be isolated and purified. Second, the purified plasmids will be digested with endonucleases to produce linear fragments. These fragments and the original circular DNA will be characterized by electrophoresis on an agarose gel.

1. Isolation and Purification of Plasmid DNA

There are several methods in isolation and purification of plasmid based on the difference between DNA of plasmid and chromosome, including three basic steps: Bacteria culture and plasmid amplification. Harvest and lysis of bacteria. Purification of plasimd.

Boiling - method is used in this exercise. Bacterial cells lysed with enzymes and detergents release plasmid and chromosomal DNA into the medium as a viscous mass. Lysozyme and detergent (for example, SDS or Triton) break the peptidoglycan layer and the membrane. Boiling water bath can precipitate protein. Plasmid DNA can be separated from most of the chromosomal DNA and other cell materials (proteins, high molecular RNA) by centrifugation. Plasmid DNA remains in the supernatant solution while high molecular weight materials are sedimented. The final purification step utilizes ethidium bromide - cesium chloride buoyant density centrifugation. Plasmids take up less ethidium bromide than chromosomal DNA macromolecules, which helps in the separation so that the DNA, either chromosomal or plasmid, can be recovered in essentially pure form. In this exercise, the final step in the purification procedure as outlined above will be omitted since it is not well sited for a classroom laboratory exercise.

2. Characterization of Plasmid DNA by Electrophoretic Mobility on An Agarose Gel

Plasmid DNA can be detected by electrophoresis on an agarose gel. DNA migrates from negative electrode to positive electrode when the pH is higher than its isoelectric point. DNA samples are mixed with a tracking dye and put into the wells on the agarose gel. Migration of DNA through the gel begins when a voltage is applied. Small fragments of DNA migrate faster than larger fragments, and covalently closed molecules migrate faster than linear and open - circular molecules. At the completion of the electrophoresis run, the gel is removed from the apparatus and stained with ethidium bromide. Ethidium bromide is an intercalating agent that is fluorescent with ultraviolet illumination, allowing visualization of DNA bands.

3. Transformation

In transformation, a free DNA molecule is transferred from a donor to a recipient bacterium. To take up DNA, a recipient cell must be competent. Natural transformation occurs in relatively few bacterial genera.

Artificial transformation is carried out in the laboratory by a variety of techniques, including treatment of the cells with calcium chloride, which renders their membranes more permeable to DNA. This approach succeeds even with species that are not naturally competent, such as *E. coli*. Relatively high concentrations of DNA, higher than would normally be present in nature, are used to increase transformation frequency.

When linear DNA fragments are to be used in transformation, *E. coli* usually is rendered deficient in one or more exonuclease activities to protect the transforming fragments. It is even easier to transform bacteria with plasmid DNA since plasmid is not as easily degraded as linear fragments and can replicate within the host. This is a common method for introducing recombinant DNA into bacterial cells. DNA from any source can be introduced into bacteria by splicing it into a plasmid before transformation.

In this exercise, plasmid pBR322 is used to transform *E. coli* K12 802. The transformant acquired resistance to tetracycline and ampicillin.

Materials and Apparatus

1. Strains

E. coli C600 (pBR322), overnight agar slant culture.

Recipient strain: *E. coli* K12 802.

Donor: plasmid pBR322.

2. Media

LB broth, M9 broth, nutrient agar.

3. Reagents

STET solution, lysozyme solution, TE solution, 1 mmol/L EDTA solution, 10 μg/ml ethidium bromide solution, tracking dye, electrophoresis buffer, 0.7% agarose gel, 60 mmol/L $CaCl_2$ solution, 4 mg/ml ampicillin solution, 5 mg/ml tetracycline, 0.9% NaCl solution.

4. Apparatus

Eppendorf (EP) tubes, microcentrifuge, UV transilluminator, horizontal slab gel electrophoresis apparatus with a power source, water bath, shaking incubator, centrifuge, pipette, petri dish, spreader.

Procedures

1. Isolation and Purification of Plasmid DNA

(1) Transfer a loopful of the culture into a sterile EP tube containing 100μl STET solution. Add 8μl of lysozyme solution and place at room temperature for 3 min. Place into boiling water bath of 100℃ for 60 s. Then place quickly into ice – water bath of 0℃ for 5 min.

(2) Centrifuge at 8000 rpm in a microcentrifuge for 5 min. Most of the chromosomal DNA and cell debris will precipitate as a white viscous clot. Pick the pellet up with a sterile tooth – stick and discard.

(3) Add 100μl of cold ethanol to the tube and mix well. Place the tube at –20℃ for 30 min. Centrifuge at 10000 rpm in a microcentrifuge for 5 min to precipitate plas-

mid DNA.

(4) Decant the supernatant. Dissolve the plasmid DNA pellet in 20 μl of TE solution for detection with electrophoresis.

2. Characterization of Plasmid DNA by Electrophoretic Mobility on an Agarose Gel

(1) Prepare the agarose gels for analysis: Bring 0. 7 g of agarose in 100 ml of electrophoresis buffer to a boil to dissolve the gel.

(2) The solution is then cooled to about 50℃ and poured into a slab mold with a comb for making wells at the top of the gel.

(3) After the gel has solidified, the comb is removed and the electrophoresis chambers are filled with electrophoresis buffer.

(4) Add 10μl of tracking dye to 40μl of the plasmid DNA. Mix well. Add the mixture to the slots on the agarose gel.

(5) Set up the gel so that the tracking dye moves towards the positive electrode. Apply the voltage at 80 V.

(6) Periodically observe the gel for migration of the tracking dye.

(7) After the run, turn off the power, disconnect the apparatus, and remove the gel.

(8) Stain the gel in ethidium bromide for 30 min. Visualize bands under UV illumination at wave of 254 nm.

(9) Draw the position of wells, the bands of chromosomal DNA, plasmid DNA, RNA and the tracking dye.

3. Transformation

(1) Inoculate a loopful of agar slant culture of recipient strain into 5ml of M9 broth. Incubate with shaking at 37℃ overnight.

(2) Inoculate 0. 5ml of the above culture into 250 ml flask containing 5ml of M9 broth. Incubate shaking at 37℃ for 2h. Centrifuge the culture at 3, 500 rpm for 15 min. Wash the bacteria with pre – cooled 0. 9% NaCl solution and resuspend the bacteria in 0. 5 ml of pre – cooled $CaCl_2$ solution.

(3) Transfer 0. 1ml competent bacteria to a sterile tube and add 0. 1 ml DNA diluted before. Place in ice – water bath for 30 min and then place in 42℃ water bath for 4 min. Place in ice – water bath again for one hour.

(4) Add twofold volume of M9 broth (about 0. 4ml). Incubate at 37℃ for 3h for phenotype expression.

(5) Dilute the mixture suitable and transfer 0. 1 ml onto the nutrient agar plate containing ampicillin (20μg/ml) and tetracycline (25μg/ml). Make DNA – control and recipient – control at the same time. Incubate at 37℃ overnight.

Results and Discussion

1. The results of electrophoresis: Draw the position of the wells, the bands of chromosomal DNA, plasmid DNA, RNA and the tracking dye.

2. The results of transformation.

Media / Test group	The growth condition on CM plate (containing Ap and Tc)
Recipient	
Plasmid DNA	
Transformation	

3. How to prepare the competent cell of *E. coli*?

4. Will the result be different if omitting the phenotype expression?

5. The growth curve must be determined before transformation, why?

实验三十八　细菌转导（局限性转导）

【实验目的】

1. 学习局限性转导的基本原理和方法。
2. 学习高频转导和低频转导。

【实验原理】

转导是以噬菌体作为媒介，将一个细胞（供体）的遗传物质传递给另一细胞（受体）的过程。转导分为普遍性转导和局限性转导。本实验以局限性转导为例，用λ噬菌体专一性转导半乳糖发酵基因的现象来了解转导的基本原理，熟悉转导实验的基本方法。用大肠埃希菌 K12S（λ）gal^+为供体，该菌带有原噬菌体（λ），控制半乳糖发酵的基因在细菌染色体上和（λ）紧密连锁。当此溶原菌受紫外光诱导后，原噬菌体（λ）被释放出来，其中有一定比例的噬菌体携带有邻近的半乳糖发酵基因，用它感染另一不能发酵半乳糖的受体菌，就可使受体菌转变为能发酵半乳糖的菌体。过程如下：

K12S（λ）gal^+（供体）

↓UV 照射

λgal^+

↓

K12S gal^-（受体）→K12S $gal^-/\lambda gal^+$转导子（局部杂合子）

【实验材料】

1. 菌种

供体菌大肠埃希菌 K12S（λ）gal^+，受体菌大肠埃希菌 K12S gal^-。

2. 培养基

肉汤培养基，加倍肉汤培养基，半乳糖 EMB 培养基。

3. 试剂

三氯甲烷，含 Mg^{2+} 的磷酸缓冲液。

4. 器材

离心机，离心管，吸管，毛细管，紫外灯，平皿。

【实验方法】

1. 噬菌体裂解液的制备

（1）过夜活化的供体菌斜面，取一环接种于 5ml 肉汤中，37℃摇育 6 小时。将供体菌离心集菌，3500 转/分离心 15 分钟。

（2）沉淀物加5ml含Mg^{2+}的缓冲液，混匀。

（3）吸取3ml菌液置于灭菌小平皿内，打开皿盖，紫外线照射30～50秒（15W距离30cm）。

（4）照射后，加入3ml加倍肉汤培养基，避光培养2～3小时。

（5）吸取培养物3 ml于离心管中，3500转/分，离心15分钟，吸出上清，加0.1ml三氯甲烷，剧烈振荡30秒，静置5分钟，再离心，小心把上清用无菌毛细管转移到另一试管，即得（λ）噬菌体裂解液。

2. 转导

（1）点试法

①过夜活化的受体菌K12S *gal*$^{-}$斜面，取一环接种于5ml肉汤中，37℃摇育6小时。将受体菌的6小时培养物5～10倍稀释后，铺满含半乳糖的EMB平板，并把剩余菌液吸干，置孵箱中烘干（稍开盖30分钟）。

②滴加（λ）噬菌体裂解液3～4滴于平板中央，37℃培养48小时，观察。同时分别取一环（λ）噬菌体，供、受体菌，点种于半乳糖EMB培养基上作对照，37℃培养过夜。观察结果。

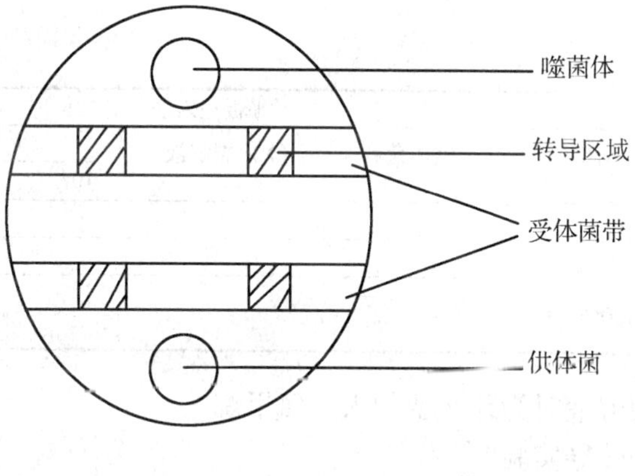

图38－1　转导点试法示意图

Fig. 38－1 Diagram of spot test for transduction

（2）涂布法

①取预先倒好的EMB平板五只，其中一只加0.1ml噬菌体裂解液，用于对照；二只分别加0.1ml经活化后的受体菌和供体菌，也用于对照；另二板加噬菌体裂解液和受体菌各0.05ml。

②用4支灭菌的涂布棒分别涂布上述各组平板，37℃培养2天，观察结果。

（3）平板注入法

①将噬菌体裂解液用肉汤稀释为10^{-1}、10^{-2}、10^{-3}、10^{-4}，从各稀释度吸取2 ml裂解液于含有2 ml经活化后的受体菌的离心管中，同时以只加噬菌体裂解液（2ml 10^{-1}裂解液加2 ml肉汤培养液）和只加受体菌（2ml经活化后的受体菌加2ml肉汤培养液）为对照。共6支灭菌离心管，37℃保温15分钟。

②取出上述6支离心管，3500转/分离心10分钟，弃上清，沉淀打匀，用生理盐水

洗涤一次，3500 转/分离心 10 分钟，弃上清，沉淀打匀，各加 2ml 生理盐水制成菌悬液。

③从上述菌悬液中各吸取 0.1 ml 于预先倒好的半乳糖基本培养基平板中，转导试验每浓度各 2 皿，供体菌、受体菌和对照各 1 皿，共 11 皿，用 7 支灭菌的涂布棒分别涂布各组平板，37℃培养 48 小时，观察结果，计数出现的转导子数并计算转导频率。

【结果与讨论】

1. 转导点试法结果。

转导实验	点试法			
	受体菌	供体菌	（λ）裂解液	受体菌 +（λ）裂解液
菌落生长情况				
菌落色泽				

2. 转导涂布法结果。

转导实验	涂布法			
	受体菌	供体菌	（λ）裂解液	受体菌 +（λ）裂解液
菌落生长情况				
菌落色泽				

3. 转导平板法结果。

转导试验	平板注入法						
	受体菌	供体菌	（λ）裂解液	转导子数/皿			
				10^{-1}	10^{-2}	10^{-3}	10^{-4}
菌落生长情况							
转导子数/ml							
转导频率 $=\frac{转导子数/ml}{噬菌体总数/ml}\times 100\%$							

4. 在制备噬菌体裂解液时为什么要加入三氯甲烷？

5. 怎样区分转化和转导实验？

38 Bacterial Transduction

Purpose

To be familiar with the principle of specialized transduction and transfer method of genes from a donor bacterium to a recipient bacterium by bacteriophage λ.

Principle

In transduction, the DNA is transferred from a donor to a recipient bacterium by a bacteriophage. For transduction, a bacteriophage must acquire a portion of the genome of the host cell in which it reproduces. There are two different types of transduction, generalized and specialized, which differ in whether they can bring about the general transfer of genes or only the transfer of specific genes. In each case, the bacteriophage acts as a carrier of bacterial genetic information.

Specialized transduction is made possible by an error in the lysogenic life cycle. When a prophage is induced to leave the host chromosome, excision is sometimes carried out improperly. The resulting phage genome contains portions of the bacterial chromosome next to the integration site. A tranducing phage genome usually is defective and lacks some part of its attachment site.

The best - studied example of specialized transduction is the lambda phage. The lambda genome inserts into the host chromosome at specific locations known as attachment or *att* sites. The *att* site for lambda is next to the *gal* and *bio* genes on the *E. coli* chromosome; consequently, specialized transducing lambda phages most often carry these bacterial genes. The lysate, or product of cell lysis, resulting from the induction of lysogenized *E. coli* contains normal phage and a few defective transducing particles. These particles are called lambda *dgal* because they carry the galactose utilization genes. Because these lysates contain only a few transducing particles, they often are called low - frequency transduction lysates (LFT lysates). Whereas the normal phage has a complete *att* site, defective transducing particles have a nonfunctional hybrid integration site that is part bacterial and part phage in origin. Integration of the defective phage chromosome does not readily take place. Transducing phages also may have lost some genes essential for reproduction. Stable transductants can arise from recombination between the phage and the bacterial chromosome because of crossovers on both sides of the *gal* site.

Defective lambda phages carrying the gal gene can integrate if there is a normal lambda phage in the same cell. The normal phage will integrate, yielding two bacterial/phage hybrid *att* sites where the defective lambda *dgal* phage can insert. It also supplies the genes missing in the defective phage. The normal phage in this instance is termed the helper phage because it aids integration and reproduction of the defective phage. These transdutants are unstable because the prophages can be induced to excise by agents such as UV radiation. Excision, however, produces a lysate containing a fairly equal mixture of defective lambda *dgal* phage and normal helper phage. Because it is very effective in transduction, the lysate is called a high – frequency transduction lysate (HFT lysate). Reinfection of bacteria with this mixture will result in the generation of considerably more transductants. LFT lysates and those produced by generalized transduction have one transducing particle in 10^5 or 10^6 phages; HFT lysates contain transducing particles with a frequency of about 0.1 to 0.5.

Materials and Apparatus

1. Strains

Donor: *E. coli* K12S (λ) gal^+, Recipient: *E. coli* K12Sgal^-.

2. Media

Nutrient broth, 2 × nutrient broth, galactose EMB agar.

3. Reagents

PBS containing Mg^{2+}, chloroform.

4. Apparatus

Centrifuge, centrifuge tubes, pipette, ultraviolet lamp, petri dishes.

Procedures

A. Preparation of Transduction Lysate

1. Transfer a loopful of overnight agar slant culture of the donor strain and inoculate into 5 ml of broth. Incubate with shaking at 37℃ for 6h.

2. Harvest the bacteria by centrifugation at 3500 rpm for 15min, resuspend them in 5 ml PBS containing Mg^{2+} and mix well.

3. Pipette 3ml the resuspended culture to a small sterile petri dish. Irradiate for 30 to 50s with the cover removed from the petri dish at a distance of 15cm from the UV lamp.

4. After irradiation, add 3ml of broth and incubate from light for 2 to 3h.

5. Transfer 3ml of the culture to a centrifuge tube and centrifuge at 3500 rpm for

15min. Pipette the supernatant liquid and add 0. 1ml of chloroform. Agitate the tube for 5min. Centrifuge again. Transfer carefully the supernatant into another tube. This is the λ lysate for transduction.

B. Bacterial Transduction

1. Spot method

(1) Transfer a loopful of overnight agar slant culture of the recipient strain and inoculate into 5 ml of broth. Incubate with shaking at 37℃ for 6h. Dilute this culture 5 to 10 fold and spread it on the surface of galactose EMB plate to form two bands and dry.

(2) Add 4 drops of λ lysate for transduction on the recipient area. Inoculate a loopful of donor culture onto the plate as donor – control. And add a drop of λ lysate for transduction onto the plate as phage – control. Incubate at 37℃ for 48h.

2. Spread method

(1) Prepare 5 plates of galactose EMB agar. Add 0. 1ml of λ lysate for transduction onto the first plate as phage – control. Add 0. 1ml donor culture onto the second plate as donor – control. Add 0. 1 ml recipient culture onto the third plate as recipient – control. Add 0. 05ml of λ lysate for transduction and 0. 05 ml of recipient culture onto each of the other two plates.

(2) Spread them out with 4 sterile glass spreaders, respectively. Incubate at 37℃ for 48h.

3. Plate method

(1) Prepare a ten – fold serial dilution of the λ lysate for transduction. Pipette 2 ml from each dilution of 10^{-1} to 10^{-4} to centrifuge tubes each containing 2 ml of recipient broth culture. At the same time make the phage – control tube and the recipient – control tube.

(2) Keep these six centrifuge tubes at 37℃ for 15 min. Centrifuge at 3500 rpm for 10min. Discard the supernatant and resuspend the bacteria. Wash with 0. 9% NaCl. Centrifuge again and resuspend the bacteria with 2 ml of 0. 9% NaCl.

(3) Pipette from each suspension onto the galactose EMB plates and spread out with glass spreaders. 2 plates for each concentration of transduction. 1 plate for each control. Incubate at 37℃ for 48h. After incubation, count the colonies of transductants and calculate transduction rate.

Results and Discussion

1. The results of spot test.

Transduction	Spot test			
	Recipient	Donor	(λ) lysate	Recipient + (λ) lysate
Growth of colonies				
Color of colonies				

2. The results of spread test.

Transduction	Spread test			
	Recipient	Donor	(λ) lysate	Recipient + (λ) lysate
Growth of colonies				
Color of colonies				

3. The results of plate test.

Transduction	Plate test						
	Recipient	Donor	(λ) lysate	Number of transductant/plate			
Growth of colonies				10^{-1}	10^{-2}	10^{-3}	10^{-4}
Color of colonies							
Transduction							
Transduction ratio = $\frac{\text{Number of transductant/ml}}{\text{Number of phage/ml}} \times 100\%$							

4. Why does chloroform be added in preparation of the λ lysate for transduction?

5. How can you differentiate experimentally between transformation and transduction?

实验三十九 枯草芽孢杆菌原生质体形成和再生

【实验目的】

学习细菌原生质体制备和再生的基本原理和方法。

【实验原理】

微生物原生质体的研究发展迅速，原生质体技术包括原生质体的制备、再生、融合、诱变、再生育种、转化和固定化等。

原生质体技术的关键是对细胞壁进行消化，以形成大量的原生质体，并使原生质体能高频率的再生细胞壁，回复到正常细胞状态。当细胞壁去除后，原生质体具有以下特征：一是无细胞壁障碍，可对膜和细胞器进行基础研究，同时可进行遗传操作；二是具有全能性，能在人工条件控制下，进行大量快速繁殖；三是可诱导融合，形成杂种细胞，为体细胞杂交提供实验材料。因此，原生质体技术无论在理论上或实践上都日益受到重视，在基因定位、酶的定位、细胞壁的合成和微生物遗传育种等工作中应用相当广泛。

通常用溶菌酶消除坚固的细菌细胞壁，细胞在高渗液中形成原生质体，然后在高渗的丰富的培养基保障原生质体再生细胞壁。由于缺乏细胞壁的保护，原生质体对外界的渗透压十分敏感，制备好的原生质体必须保存在高渗溶液中，而在低渗条件（如水）或SDS作用下，即可破裂。利用此法即可计算原生质体形成率及再生率。

【实验材料】

1. 菌种

枯草芽孢杆菌168。

2. 培养基

LB 培养基（液体和固体）；HM 再生培养基：LB 培养基加 0.5mol/L 蔗糖和 20 mmol/L $MgCl_2 \cdot 6H_2O$；HM 半固体培养基。

3. 试剂

SMM 溶液：0.5mol/L 蔗糖，0.02mol/L 顺丁烯二酸盐，0.02mol/L $MgCl_2 \cdot 6H_2O$，用 NaOH 调 pH 为6.5；SMMP 液：5 倍浓 LB 液体培养基2 份，SMM 液7.5 份，2% 牛血清白蛋白0.5 份（用膜滤器除菌）；溶菌酶液：5mg/ml 用 SMMP 配制。

4. 器材

恒温水浴，离心机，吸管，平皿，试管等。

【实验方法】

1. 原生质体形成

（1）取斜面菌少量，接种至5ml LB液体培养基内，37℃，180转/分摇育过夜。

（2）将菌悬液按2%接种量接种至装有40ml LB液体培养基的三角瓶中，继续摇育约3小时，至OD_{570}=0.4~0.6。

（3）吸取菌悬液10ml，3500转/分，离心15分钟，弃上清，沉淀用SMM洗一次，重新悬浮于1ml SMMP液。

（4）吸0.1ml菌悬液加入至0.9ml生理盐水的试管中，10倍系列稀释至10^{-6}，取10^{-6}稀释度的菌液0.1ml，加入平皿，随即倾注融化并冷至50℃的LB培养基约10ml，摇匀待凝，做2个平板，37℃培养过夜，观察结果，并计数。

（5）余下0.9ml，加溶菌酶0.1ml，使终浓度为0.5mg/ml，保温45分钟，镜检，几乎100%形成原生质体，由杆状变成球状。

（6）随即3500转/分离心10分钟，沉淀悬浮于原体积（0.9ml的SMMP液），即得原生质体悬液。

2. 原生质体再生

（1）吸取0.2ml原生质体悬液，加入1.8ml无菌水的试管中，作10倍稀释至10^{-2}。放置30分钟，使原生质体破裂，取适当稀释度的菌液0.1ml，加入平皿中，倾注约10ml的LB培养基，摇匀，待凝，37培养24小时，计数。

（2）吸取0.2ml原生质体悬液，用SMMP液10倍稀释至10^{-4}~10^{-5}。用双层平板法（先在平皿中倒入HM培养基15ml作为底层，再取适当稀释度菌液0.1ml，加入装有5ml冷至48℃的HM半固体培养基试管中，迅速摇匀，倾注底层平板上），经37℃培养48小时，计数。

【结果与讨论】

1. 将实验结果填入下表。

酶作用前		酶作用后						形成率	再生率
稀释度		水稀释			SMMP稀释			$\frac{A-B}{A}\times 100\%$	$\frac{C-A}{A-B}\times 100\%$
10^{-6}（菌数/皿）	原液（个/ml）A	10^{-1}（菌数/皿）	10^{-2}（菌数/皿）	原液（个/ml）B	10^{-4}（菌数/皿）	10^{-5}（菌数/皿）	原液（个/ml）C		

注：A：酶作用前的菌数/ml＝平均数/皿/加菌量×稀释倍数

B：酶作用后的非原生质体数/ml

C：酶作用后原生质体数/ml＋非原生质体数/ml

2. 在制备原生质体为什么要加入蔗糖？

39 Formation and Regeneration of *Bacillus subtilis* Protoplast

Purpose

To be familiar with the principles and methods for the formation and regeneration of bacterial protoplast.

Principle

Protoplasts technology includes protoplast preparation, regeneration, fusion, mutagenesis, breeding, transformation and immobilization.

The key of protoplast technology is digestion of the cell wall to form many protoplasts, and protoplast regeneration in high frequency, back to the status of normal cells. When the cell wall removed, the protoplast are with the following characteristics: First, cell - based research on membrane and organelle and genetic manipulation may be carried out without cell wall barriers; Second: Protoplasts are totipotent and could massive rapid product in artificial conditions; Third: Protoplast fusion could be induced to form a hybrid cell for somatic cell hybridization. Therefore, the protoplast technology is more and more important in theory or practice, such as gene mapping, targeting the enzyme, cell wall synthesis and microbial genetics and breeding.

Usually lysozyme is used to eliminate the solid bacterial cell wall. Fomation and regeneration of protoplasts are at hypertonic condition. Lack of cell wall protection, protoplast is very sensitive to outside osmotic pressure and must be well preserved in hypertonic solution. Protoplast will rupture in hypotonic conditions (such as water) or SDS. So the rate of protoplast formation and regeneration can be calculated.

Materials and Apparatus

1. Strains

Bacillus subtilis 168

2. Media

LB broth and agar; HM regeneration medium: LB agar with 0.5mol/L sucrose and 20mmol/L $MgCl_2 \cdot 6H_2O$; HM regeneration medium (semi - solid).

3. Reagents

SMM Solution: 0.5mol/L sucrose, 0.02mol/L maleic acid, 0.02mol/L $MgCl_2$ ·

$6H_2O$, adjusted pH with NaOH to 6.5; SMMP solution: 5 - fold concentrated LB broth: SMM solution: 2% bovine serum albumin = 2 : 7.5 : 0.5, with membrane filter sterilization; lysozyme solution: 5mg/ml prepared with SMMP solution.

4. Apparatus

Constant temperature water bath, centrifuge, pipette, plate, tube and so on.

Procedures

1. Formation of Protoplasts

(1) Inoculate a loopful of agar slant culture to 5ml LB broth and incubate at 37℃, 180 rpm/min.

(2) Inoculate the above culture into flask containing 40ml of LB broth. Incubate shaking at 37℃ for 3h (OD_{570} = 0.4 ~ 0.6).

(3) Pipette 10ml of the culture and centrifuge at 3500rpm for 15 min. Discard the supernatant Wash the sediment with SMM solution and resuspend in 1ml SMMP solution.

(4) Pipette 0.1ml bacterial suspension to 0.9% NaCl solution, 10 - fold serial dilute to 10^{-6}. Pipette 0.1ml of to petri dish and pour melted and cooled to 50℃ LB agar about 10ml and mix. Make 2 plates. Observe and count the colonies after 37℃ overnight cultivation.

(5) Add lysozyme solution 0.1ml to the remaining 0.9ml to make the final concentration of 0.5mg/ml. Incubate 45min and then examine under microscope. Almost 100% formation of protoplasts with shape change from rod to ball.

(6) Centrifuge at 3500 rpm for 10min. Resuspend the sediment in SMMP solution 0.9ml to obtain the protoplasts suspension.

2. Regeneration of Protoplasts

(1) Pipette the protoplasts suspension 0.2ml to 1.8ml sterile water, 10 - fold dilute to 10^{-2}. Lay for 30min. So the protoplasts rupture. Pipette appropriate dilution bacteria 0.1ml to petri dish and pour about 10ml LB agar. Incubate at 37℃ for 24h and count.

(2) Pipette the protoplasts suspension 0.2ml to 1.8ml SMMP solution, 10 - fold dilute to 10^{-4} ~ 10^{-5}.

Pour 15ml HM regeneration agar to petri dish as lower layer. Pipette appropriate dilution bacteria 0.1ml to the tube containing 5ml HM semi - solid regeneration medium. Mix quickly and pour to the lower layer. Incubate at 37℃ for 48h and count.

Results and Discussion

1. The results of protoplasts formation and regeneration.

Before action of lysozyme		After action of lysozyme						Formation rate	Regeneration rate
Dilution degree		Dilute with water			Dilute with SMMP			$\frac{A-B}{A}\times 100\%$	$\frac{C-A}{A-B}\times 100\%$
10^{-6} (CFU/plate)	original bacteria (CFU/ml) A	10^{-1} (CFU/plate)	10^{-2} (CFU/plate)	original bacteria (CFU/ml) B	10^{-4} (CFU/plate)	10^{-5} (CFU/plate)	original bacteria (CFU/ml) C		

注：A： CFU/ml before action of lysozyme = average/plate/bacterial suspension × dilution fold

B：non – protoplasts after action of lysozyme /ml

C：protoplasts after action of lysozyme /ml + non – protoplasts after action of lysozyme /ml

2. Why should sucrose be used in protoplasts formation and regeneration?

第七部分　免疫实验

实验四十　抗原与免疫血清的制备

将抗原注射于动物体内，使其产生相应的抗体。待动物血清中产生大量的抗体时，采出动物的血液，分离析出血清，此即含抗体的抗血清或称免疫血清。制备特异性强、效价高的免疫血清对于微生物的鉴定，疾病的诊断与治疗以及抗原的分析均有很大用途。

【实验目的】

掌握抗原与抗体的制备方法，并为相关的免疫学实验准备材料。

【实验原理】

动物产生抗体的量，一方面可因动物的种类、年龄、营养状况及刺激部位的感受性的不同而不同，另一方面还与抗原的种类、注射量、注射途径、注射次数以及注射的间隔时间有关。抗原量低到一定限度时，抗体不能产生或用现有方法不能测出，但超过最高限度对抗体的产生反起抑制作用。在最低限度以上，抗体根据抗原量的增加而增加，达到最高限度时，即不再增加。每种类型的抗原，其使用量都是根据经验而得出的。

本实验的抗原是细菌和血细胞，采用每毫升中含 9 亿个细菌及 10% 绵羊红细胞的抗原浓度，由于抗原的免疫原性较强，不需加弗氏佐剂。抗原初次注射后，经一段诱导期，血清内即可找到抗体，以后逐渐上升，抗体总量一般不高，然后逐渐下降。但再次注射时，抗体总量迅速上升到最高水平（高于初次注射），而且维持时间也长。因此，制备抗体一般需要多次注射抗原，才能得到高效价的免疫血清。最常用的注射途径是静脉注射。

针对细菌的免疫血清称为抗菌血清，针对红细胞的免疫血清称为溶血素。

【实验材料】

1. 试剂

24 小时培养的大肠杆菌斜面，硫柳汞，0.5% 石炭酸生理盐水，20% 绵羊红细胞悬液（由羊颈静脉以无菌手续取血，置于灭菌的含有玻璃珠之锥形瓶中，振摇去其纤维，使其不凝固，并用生理盐水洗涤 3 次，配成绵羊红细胞悬液）。

2. 其他

2kg 左右的健康雄家兔或未孕的健康雌家兔，McFarland 比浊管，酒精棉花，碘酒棉花，消毒干棉花，灭菌吸管，毛细滴管，试管，离心管，2ml 和 20ml 注射器针头，灭菌细口瓶，离心机等。

【实验步骤】

（一）抗菌免疫血清的制备

1. 抗原的制备

（1）吸取灭菌的0.5%石炭酸生理盐水5ml，注入大肠埃希菌斜面培养物上，将菌苔洗下。

（2）用无菌毛细滴管吸取洗下的菌液，注入无菌小试管。

（3）将此含有菌液的小试管放60℃的水浴箱中1小时，并不时摇动。

（4）与比浊管对照，比浊，判定含菌液试管中细菌的含量。

（5）用0.5%石炭酸生理盐水将菌液稀释至9×10^8 CFU/ml。

（6）无菌试验　将已稀释好的菌悬液接种少量于肉汤培养基内，培养24～48小时，观察有无细菌生长，如无细菌生长，即可放冰箱备用。

2. 免疫血清的制备

（1）注射动物　消毒注射器与针头，用消毒注射器抽取以上制备好的大肠埃希菌抗原（抽取前摇匀），按表40－1所列剂量与日程注射家兔耳静脉。第14日，自耳静脉采血lml，分离血清，测其凝集效价，如合格即可大量采血。第16日采血。

表40－1　免疫家兔的抗原注射量与时间

日程	菌液注射剂量（ml）
第1日	0.2
第2日	0.4
第3日	0.6
第4日	0.8

（2）家兔心脏采血　助手的左手紧攫家兔的颈部及两只前腿，右手握家兔后肢，使家兔仰卧，胸腹部向上，固定。或将家兔仰卧手术台上，用绳缚其四肢而固定。

用左手在家兔肋骨左侧探得心脏搏动最剧烈处剪毛，然后分别用碘酒棉花与酒精棉花消毒此部位，将右手所执注射器（注射器为20ml，针头长约35～40mm，可作采家兔心血之用）从心脏搏动最剧烈的肋骨间隙刺入心脏，若针刺准确，则此时心血涌入注射器中，徐徐抽取血液，达足量（一般每兔可抽20ml而不影响存活）时迅速拔出注射器，针刺处按以清洁干棉球，并立即将所得的血液以无菌手续注入一已灭菌的大试管或平皿内，37℃尽量放成最大斜面，凝固后放入4～6℃的冰箱中，使其自然析出血清。

（3）制备血清　自（2）制得凝固血液的容器中，用已灭菌的毛细滴管分出血清，若血清中带有红细胞，则须离心沉淀以去掉红细胞。然后将血清装入灭菌细口瓶中，并测定抗血清的效价（见实验四十三凝集反应实验）。加入防腐剂，使血清含有1/10000硫柳汞。用蜡封瓶口，贴上标签，注明抗血清名称、凝集效价及日期，放冰箱备用。

（二）溶血素的制备

采用静脉注射方法，在第1、6、11、16天每天注射3ml绵羊红细胞悬液，再于第

19～21 天取血测定其中抗体含量，如效价在 1∶16 以上即达到要求，应及时采血。

【结果与讨论】

1. 制备大肠埃希菌抗原时，使用石炭酸生理盐水和 60℃加热 1 小时的目的是什么？
2. 在动物体内制备免疫血清为什么要多次注射？

PART SEVEN Experiments for Immunology

40 Preparation of Antigen and Antiserum

Injecting antigen into animal will induce antibody. After a large quantity of antibody have been produced in animal serum, collect animal blood and separate out this serum. This serum is called antiserum or immunoserum and is used widely in microorganism identity, disease diagnose and treatment, antigen analysis.

Purpose

To be familiar with the preparation of antigens and antibodies.

Principle

The quantities of antibodies is not only depend on animal species, age, nutrition and injecting position, but also depend on the kinds, quantities of antigens and the injection ways, times and interval. The quantities of antibody will be raised as the quantities of antigen raised. Each usage of antigen is depend of experiments.

The antigens in this exercise is bacteria and blood cells, with the concentration of 9 10^8 CFU/ml and 10% red blood cell. Because the immunogenicity is powerful, it doesn't need Freund adjuvant. In primary injection, after a period of inducement, the antibody in serum could be found. The quantities of antibodies will rose gradually, but the total amount is not high, then it will decline. In secondary injection, the total amount of antibody will rise to the highest point quickly and keep it there for a long time. So we need to inject rabbit more times in order to get higher titer of immunoserun. The usual way for injecting antigen is intravenous injection.

The immunoserum for bacteria is called anti – bacterial serum. The immunoserum for red blood cell is called haemolysin.

Materials and Apparatus

1. Reagent

The agar slant of *E. coli* (cultured for 24h), thimerosal, 0.5% carbolic acid saline, 20% sheep red blood cell, broth culture.

2. Others

Healthy male rabbit or healthy female rabbit, weight 2 kg. McFarland turbidity compare tubes, alcohol – saturated cotton applicators, cotton applicators saturated with tincture of iodine, sterile dry cotton swab, sterile pipette, capillary tube pipettes with pipettor, tubes, tubes for centrifugate, 2ml and 20ml syringes and needles, sterile bottles, sterile Petri dishes, centrifuge, 60℃ water bath.

Procedures

1. Preparation of anti – bacterial serum

(1) Prepare antigens

a. Add sterile 0.5% carbolic acid saline 5ml into the agar slant of *E. coli* to suspend bacteria.

b. Transfer the suspension with sterile capillary tube pipettes to sterile tubes.

c. Place tubes into 60℃ water bath for 1 hour. Shake frequently.

d. Compare tubes turbidity with McFarland turbidity.

e. Compare tubes to measure concentrations of bacteria in tubes.

f. Dilute bacteria with sterile 0.5% carbolic acid saline into 900 000 000cfu/ml.

g. Culture some diluted bacterial suspension into broth culture for 24 ~48 hours. If no growth was observed, store the diluted bacteria into refrigerator.

(2) Prepare immunoserum

a. Inject animals Sterilize syringes and needles. Absorb the diluted bacterial suspension, inject into rabbit ear vein (Table 40 – 1). On the fourteenth day, take 1 ml from rabbit ear vein, separate serum and determine its agglutination titer. If the titer reaches standard, gather large quantity of serum and, gather serum on the sixteenth day.

Table 40 – 1 The injection quantities of antigen and times in immune rabbit.

Time	The injection quantity of bacterial suspension (ml)
The first day	0.2
The second day	0.4
The third day	0.6
The forth day	0.8

b. Gather serum from rabbit heart Let the assistant grasp rabbit neck and two forelegs with left hand, grasp rabbit hind leg with right hand to make rabbit lie supine, then fix it. Or make rabbit lie supine on operating table, fix its four limbs with ropes.

Look for a place in rabbit left rib which its heart beat most violently. Cut feather and sterilize with cotton applicators saturated with tincture of iodine and alcohol – saturated cotton applicators. Stab into rabbit heart through ribs interval with 20ml syringe. If stab accurate, the heart blood will gush into syringe. Draw the syringe slowly to get enough blood (a rabbit can keep alive after missing 20 ml blood) . Then pull out the syringe quickly and press on the stab place with sterile dry cotton swab. Add rabbit blood into a sterile Petris dish with sterile operation. Make the biggest blood slope under 37℃. After stiff, put it under 4 ~ 6℃ to let serum separate naturally.

c. Prepare the serum Transfer the serum with sterile capillary into sterile bottles. If there are red blood cells in the serum, centrifuge to get rid of it. Determine the titer of antiserum (reference the method in agglutination reaction) . Add thimerosal into serum. The concentration of thimerosal is 1/10000. Seal the bottle with wax. Paste labels that indicate name, titer and date. Store it in refrigerator.

2. Preparation of haemolysin

Inject 3ml 10% sheep red blood cell in the first, sixth, eleventh, sixth day with intravenous injection. At 19 ~ 21 day, obtain small quantity of rabbit blood and determine the titer. If the titer is higher than 1 : 16, gather blood as method mentioned above.

Results and Discussion

1. Why use carbolic acid saline and heat in 60℃ for 1h in preparing *E. coli*?
2. Why it need more times for injecting antibody to get immunoserum in animal?

实验四十一 吞噬细胞功能的测定

固有性免疫是机体针对异物的天然防御功能，主要包括吞噬作用、生理屏障、体液因子作用和自然杀伤作用等。其中起吞噬作用的细胞主要包括单核－巨噬细胞系统和中性粒细胞。它们具备了吞噬和消化病原体、细胞碎片的能力。在固有免疫中扮演重要角色。

I 中性粒细胞的吞噬作用

【实验目的】

了解中性粒细胞的吞噬功能。

【实验原理】

在体外，将葡萄球菌与新鲜血液混合，血液中的中性粒细胞将吞噬球菌。结果可通过染色观察到。因此可通过测定吞噬细胞的吞噬功能，来了解机体的免疫状态或判断药物对机体免疫功能的影响。

【实验材料】

1. 菌种

葡萄球菌培养物。

2. 其他

新鲜血液，肝素抗凝剂，无菌生理盐水，玻片，滴管，载玻片，Wright 染液，无菌水，明视野显微镜。

【实验步骤】

1. 将试验菌种（葡萄球菌）接种于适宜的培养基中，培养，以比浊法调整菌液浓度，至约 5×10^8 个菌/ml，加热 100℃、15 分钟杀菌。保存冰箱待用。（实验前已准备好）

2. 取新鲜血液（人指尖或动物）2～3 滴置凹玻片中，加一滴肝素混匀抗凝（也可用 3.8% 枸橼酸钠），然后加菌液 2～3 滴，混匀后放于潮湿的容器中，37℃保温 20～30 分钟，每 10 分钟摇动一次。

3. 保温后，立即取悬液一滴，在玻片上推成薄膜（图 41－1），晾干，加瑞氏（Wright）染液，1 分钟后加等量的蒸馏水，吹匀，5～10 分钟后弃去染液，水冲，吸水纸吸干，油镜镜检。

4. 镜检视野中有大量红细胞存在，被染成深蓝色的、颗粒极小的葡萄球菌散在于细

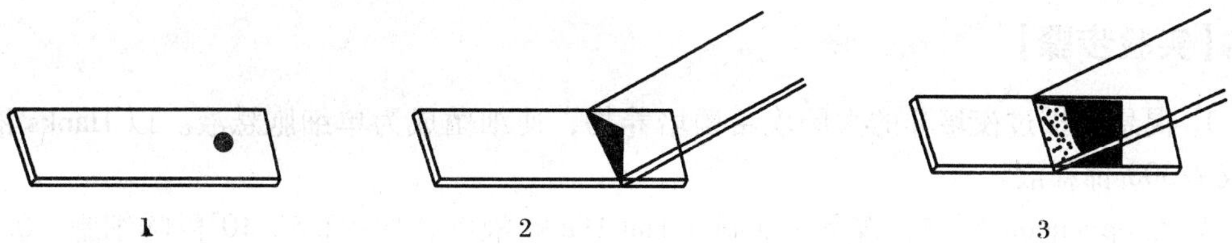

图 41 -1　推片制作示意图

Fig. 41 -1 Preparation of sample

胞间。寻找不同视野，可发现蓝色无细胞浆的小淋巴细胞和紫红色的中性粒细胞（细胞核为分叶状），中性粒细胞中可找到许多葡萄球菌。

5. 如欲了解机体免疫功能的强弱，可以观察 100 个吞噬细胞，计算吞噬百分数或吞噬指数：

吞噬百分数 = 吞噬了细菌的吞噬细胞数/观察到的吞噬细胞总数

吞噬指数 = 被吞噬的细菌总数/观察到的吞噬细胞总数

正常参考值，吞噬率在 61.39% ~64.11% 之间，吞噬指数在 1.009 ~1.107 之间。

【结果与讨论】

1. 记录实验结果，作图描述中性粒细胞的吞噬作用，计算吞噬指数和吞噬率。
2. 为什么吞噬细胞在机体免疫应答过程中起重要作用?

Ⅱ　巨噬细胞杀菌功能的检测

【实验目的】

了解巨噬细胞的杀菌作用。

【实验原理】

巨噬细胞具有吞噬和杀伤细菌的作用。这一作用可通过巨噬细胞吞噬细菌的数量与孵育较长时间后细菌数量的比较来确定。

【实验材料】

1. 菌种

大肠埃希菌培养物。

2. 试剂

Hanks 液、血清（临用前融化冰浴）、Hanks 液/5% 血清（冰浴）、营养琼脂培养基、无菌水。

3. 其他

外周血分离的单核细胞、1.5ml eppendorf 管、振荡器、10ml 无菌玻璃离心管（配螺口塞）。

【实验步骤】

1. 混悬振摇过夜培养的大肠埃希菌培养物，使细菌成为单细胞悬液。以 Hanks 液制备成 1/300 稀释液。

2. 在 eppendorf 管中，按下文配制 0.1ml Hanks 液混合物：2.5×10^6巨噬细胞、0.3ml 经过振摇的细菌培养物稀释液（菌浓度 2.5×10^6/ml）、50μl 冰预冷的血清、加 Hanks 液至 1ml。密封管盖。

3. 将小管在 37℃振荡 15 ~ 20 分钟。洗去胞外多余细菌。用 1ml Hanks 液/5% 血清重悬细胞。

4. 在每个玻璃离心管中加入 0.9ml 无菌水。以 0.1ml 细胞混合物加入第一管，开始进行 10 倍的系列稀释。每一稀释度均需更换吸管并轻微振荡。

5. 振荡混匀后从每管中取出 0.1ml，铺入 37℃预温的琼脂平板中。

6. 同时，盖紧 eppendorf 管，置于 37℃，孵育 90 ~ 120 分钟。

7. 试管置冰上以使细菌生长停止。按照 4 ~ 5 步的方法制备系列稀释液，并铺入平板。

8. 37℃培养 24 ~ 48 小时。计数并比较菌落数。

【结果与讨论】

记录巨噬细胞杀伤细菌的实验结果。

41 Assay of the Function of Phagocytes

Innate immunity refers to the inborn ability that can resist a foreign matter. It mainly include phagocytosis, physiologic barriers, humoral proteins effect and natural killing etc.. Phagocytes mainly include mononuclear - phagocyte system and neutrophils, which can engulf and digest pathogen, cellular debris *etc.* They play an importment role in innate immunity.

I The phagocytes of neutrophils

Purpose

To realize the function of neutrophils.

Principle

In vitro, mixed fresh blood with *staphylococcus*, neutrophils in blood will engulf coccus. The results can be observed by staining. Through these results, We can realize the immunity state of a body or the effect of medicines to body's immunocompetence.

Materials and Apparatus

1. Culture

Culture of *Staphylococcus.*

2. Apparatus

Fresh blood, heparin, sterile physiological water, slide, Wright's stain, bright - field microscope, pipettes, sterile water.

Procedures

1. Prepare the cultures of *Staphylococcus* with 5×10^8 CFU/ml. Store at 4℃ after treated under 100℃ for 15min.

2. Drip 2 ~ 3 drops of fresh blood on the slide, mixed it with a drop of heparin. Then add 2 ~ 3 drops of coccus into blood, mix it and put it in wet box for 20 ~ 30min in 37℃, wave the slide every 10min.

3. Drag a drop of sample over a surface of a slide forming a continuous smear (Fig. 41.1). Stain it with Wright's stain for 1 min, then add the equal quantity of water, discard the fluid on the slide after 5 ~ 10min, gently wash and air dry the slide.

Observe the slide under oil - immersion lens.

4. In fields of vision, there are mass of red blood cell and small dark blue particles (cocci). Search many fields, you will find lymphocytes, which are blue and have no cytoplasm, and neutrophils, which are purplish red and have a multilobed nucleus. There are many cocci in neutrophils.

5. The percentage of phagocytosis and the index of it are related to the degree of body immunity. You may observe 100 phagocytes, and calculate as follows:

(1) Percentage of phagocytosis = the number of phagocytes that engulf bacteria / the total number of phagocytes.

(2) Index of phagocytosis = the number of engulfed bacteria / the total number of phagocytes.

The normal reference of percentage of phagocytosis is between 61.39% ~ 64.11%, The normal reference of index of phagocytosis is between 1.009 ~ 1.107.

Results and Discussion

1. Draw the diagram of phagocytosis and calculate the percentage and index of phagocytosis.

2. Why are phagocytes so important in the immune response?

Ⅱ Macrophage killing of bacteria

Purpose

To realize the killing function of macrophage.

Principle

Macrophages can engulf bacteria and killed them. This ability is measured by quantitating cell - associated bacteria after a brief phagocytosis period, then determining how many organisms remain following a longer incubation.

Materials and Apparatus

1. Culture

Culture of *E. coli.*

2. Reagent

Hanks' Balanced Salt Solution, normal serum, freshly thawed and kept on ice, Hanks solution /5% normal serum, ice - cold, broth agar medium, sterile water

3. Others

Monocytes from human PBMC, 1.5ml eppendorf tubes, Labquake shaker, 10ml sterile glass tubes with screw caps.

Procedures

1. Vortex overnight culture of *E. coli* and prepare a 1/300 dilution in Hanks solution. Before removing it, vortex the cultures to ensure that bacteria are added as single cells.

2. In a eppendorf tube, mix the following: 2.5×10^6 macrophages in 0.1ml Hanks' Solution, 0.3 ml vortexed dilute bacterial culture (2.5×10^6 bacteria). 50μl ice - cold normal serum, Hanks solution to 1 ml, close caps tightly.

3. Place tubes in Labquake shaker and rotate 15 to 20min at 37℃. Wash away extracellular bacteria using a simple wash. Resuspend cells in 1ml Hanks solution /5% serum.

4. Set up four screw - cap glass tubes containing 0.9ml sterile water each. Starting with 0.1ml of cell mixture, make four 1/10 serial dilutions into these tubes, using a new pipet for each dilution and vortexing briefly between dilutions.

5. Briefly vortex each tube and plate 0.1ml, in duplicate, on broth plates prewarmed to 37℃.

6. Cap and tightly seal undiluted sample tubes. Incubate samples 90 to 120min at 37℃.

7. Place tubes on ice to stop bacterial growth. Prepare serial dilutions and plate samples as described in step 4 ~ 5.

8. Incubate 24 to 48h at 37℃. Count the colonies and compare the number of colonies at step 5 to the number of colonies at step 7.

Results and Discussion

Record the results of the macrophage killing.

实验四十二　体液因子的免疫功能测定

体液中有着多种免疫因子，如补体、溶菌酶、干扰素等。它们具有引发炎症反应、调理吞噬、介导对细胞的杀伤、裂解细菌等多重作用。

Ⅰ　补体导致的溶血反应

【实验目的】

验证补体的溶血作用。

【实验原理】

补体的活性可通过绵羊红血球、溶血素以及豚鼠来源补体间相互作用导致的溶血现象来测定。绵羊红细胞和溶血素的混合，可导致凝集现象出现。如果向系统中添加补体，溶血作用发生。溶血作用常用于测定总补体活性或溶血素的效价。

【实验材料】

绵羊红血球悬液；溶血素（经 56℃加热半小时处理）；补体（来自豚鼠血清）；10ml 玻璃试管、无菌生理盐水、无菌吸管、移液管、37℃水箱等。

【实验步骤】

1. 取小试管三只，按表 42－1 所示加各种材料于试管中。

表 42－1　补体溶血反应检测

管号	2% 羊红血球（ml）	溶血素（2 单位）	补体（2 单位）	生理盐水（ml）	结果
1	0.5	0.5	0.5	0.5	
2	0.5	0.5	－	1.0	
3	0.5	－	0.5	1.0	

2. 操作完毕后，将上述三个试管放 37℃水浴 15～30 分钟，观察有无溶血现象（如血球溶解，混浊液变为红色透明的液体）。

【结果和讨论】

根据实验结果，完成表 42－1。

Ⅱ　溶菌酶的溶菌作用

【实验目的】

学习溶菌酶活性的检测方法。

【实验原理】

溶菌酶广泛存在于机体的泪液、唾液、痰、乳汁、血清等处，在一种重要的体液免疫因子。它能特异性地裂解革兰阳性菌细胞壁肽聚糖分子间的 β－1，4 葡萄糖苷键，而对革兰阴性细菌细胞壁作用微弱。溶菌酶活性可通过检查其对指定敏感菌株裂解作用来进行测定。

【实验材料】

1. 菌种

溶壁微球菌（*Micrococcus lysodeikiticus*）菌液，大肠埃希菌菌液。

2. 其他

1% 高层琼脂（以 1g 优质琼脂溶解于 100ml、1%、15mol/L、pH6.4 磷酸缓冲液制成），无菌生理盐水，无菌平皿、无菌小滤纸片、无菌吸管、镊子。

【实验步骤】

1. 取 1% 高层琼脂一支加热融化，待冷至 50℃左右时，加入溶壁微球菌菌液 1ml，倾注平板成含菌平板。
2. 用镊子夹取滤纸片，沾以唾液（滤纸片应完全湿润），放置于含菌平板的适当区域内。
3. 同时用滤纸片沾取生理盐水，放置于同一平板另一区内，作为对照。
4. 置 37℃培养 18～24 小时，观察结果。如为定量测定，需与溶菌酶标准品对比进行实验。
5. 用大肠埃希菌菌液重复上述步骤。

【结果与讨论】

1. 记录实验结果，并将溶菌酶的溶菌作用实验结果作图。
2. 为什么溶菌酶对溶壁微球菌的效果与对大肠埃希菌的效果不同？

42 Assay of the Function of Humoral Immune Factor

There are many immune factor in humoral, such as complement, lysozyme, interferon etc. Their activities fall into a number of broad categories including production and regulation of inflammation, opsonization of foreign materials for phagocytosis, mediation of direct cytotoxicity against cells and lysing the bacteria cell wall.

Ⅰ The Hemolytic Assay of Complment

Purpose

To validate the hemolytic of complement.

Principle

Complement activity can been measured and quantitated using hemolytic assays with sheep erythrocytes, and a complement source, usually guinea pig serum. When mixing the sheep erythrocytes with hemolysin, the agglutination will be found. And if added complement, the hemolysis will happen. The hemolytic assay was used in measuring activities of complement or potency of hemolysin, *etc.*

Material

The suspended sheep erythrocytes; hemolysin (56℃ water bath 30min); complement (from guinea pig serum); 10ml glass tubes; sterile physiological saline, sterile pipettes, 37℃ water bath.

Procedure

1. Add the materials into three glass tubes according to the table 42 – 1:

Table 42 – 1 The hemolytic assay of complment

The number of tubes	2% sheep erythrocytes (ml)	hemolysin (2 unit)	complement (2 unit)	Physiological saline (ml)	phenomenon
1	0.5	0.5	0.5	0.5	
2	0.5	0.5	–	1.0	
3	0.5	–	0.5	1.0	

2. Place in water bath at 37℃ for 15 ~ 30 min. Observe the phenomenon (if the hemolysis happen, the solution will looked transparence and red).

Results and discussion

Finish the table 42 - 1.

Ⅱ Lysozyme

Purpose

To detect the potency of lysozyme in saliva.

Principle

Lysozyme is an important kind of humoral protein, which has innate immunity effect. It is a natural component of tears, saliva, phlegm, milk, serum etc. Lysozyme specifically degrade the β - 1, 4 glycosidic bond in peptidoglycan, the main component of the Gram's positive bacteria. Its effect to lyse the Gram's negative bacteria cell wall is weak.

The potency of lysozyme can be tested through determining the lysozyme's lyse effect to sensitive bacteria.

Materials and Apparatus

1. Culture

Culture of *Micrococcus lysodeikiticus*, culture of *E. coli.*

2. Apparatus

Saliva, 1% high - layer agar (1g agar dissolved in 100 ml, 15 mol/L, pH6.4 phosphate buffer), sterile Petri dish, small sterile filter paper, sterile physiological saline, sterile pipettes, sterile tweezers.

Procedures

1. Prepare 1% high - layer agar plate with *Micrococcus lysodeikiticus.*
2. Place a slice of filter paper, dampen with saliva thoroughly, on the dish, by using tweezers.
3. Place a slice of filter paper, dampen with sterile physiological saline thoroughly, on the same dish, by using tweezers.
4. Incubate the dish for 18 ~ 24h at 37℃. Observe the result. If you want to test quantity, you should set up a positive control with standard sample.
5. Repeat the same procedures with cultures of *E. coli.*

Results and Discussion

1. Draw the diagram of lyse.

2. Why the lysozyme effect to *Micrococcus lysodeikiticus* and to culture of *E. coli* is difference?

实验四十三　适应性体液免疫检测
——凝集反应和沉淀反应

抗原抗体反应又称血清学技术，包括凝集反应和沉淀反应。颗粒状抗原（如细菌、红细胞）与对应抗体在电解质（生理盐水）参与下产生大小不等的凝块，称为凝集反应。可溶性抗原（如多糖、蛋白质、类脂等）与相应的抗体混合，在适量电解质的存在下，二者比例适合即有沉淀物出现，称为沉淀反应。凝集反应和沉淀反应是经典的血清学反应，在实验室研究和临床诊断工作中应用广泛。

Ⅰ　凝集反应

【实验目的】

1. 了解凝集反应类别。
2. 掌握玻片凝集与试管凝集的操作与结果观察方法。

【基本原理】

凝集反应可分为直接凝集反应、间接凝集反应、协同凝集反应和抗人球蛋白实验等。直接凝集反应为最简便、常用的一种实验方法，在操作上可分为玻片法与试管法两种。前者多用已知免疫血清来诊断抗原（如未知的细菌），为诊断肠道传染病菌的重要手段，也用于人类血型的测定， 其突出优点是极为快速。试管法多用已知抗原来检测抗体。例如诊断伤寒、副伤寒的肥达反应便是试管凝集反应。试管法可定量。试验时将待检血清用生理盐水作稀释，然后加入等量的标准抗原混合，在一定温度下经一定时间后观察结果。血清发生沉淀反应的最大稀释倍数的倒数即为血清中抗体的效价。

【实验材料】

1. 菌种

伤寒杆菌（*Salmonella typhimurium*）培养物。

2. 试剂

伤寒杆菌 H 血清，生理盐水，诊断血清，1∶10 倍稀释的伤寒杆菌“H”血清。

3. 其他

洁净玻片，接种环，酒精灯，蜡笔，洁净用小试管（测血清），1ml 吸管，试管架等。

【实验方法】

1. 玻片凝集反应

（1）取玻片一张用蜡笔划分为两区，一端标记“S”，即伤寒杆菌。两边各放盐水一小滴。

（2）一端加一白金耳伤寒杆菌诊断血清，另一端加生理盐水对照。

（3）取伤寒杆菌一接种环加于盐水一端磨匀。另取少许菌加于有诊断血清一端中研磨均匀。

（4）轻轻摇动玻片，数分钟后，肉眼观察结果。（观察后玻片应做消毒处理）

（5）如菌体凝集成肉眼可见的颗粒团块或絮状，其周围液体澄清，则为阳性反应。生理盐水对照不应发生凝集，为阴性反应（参见图43－1）。

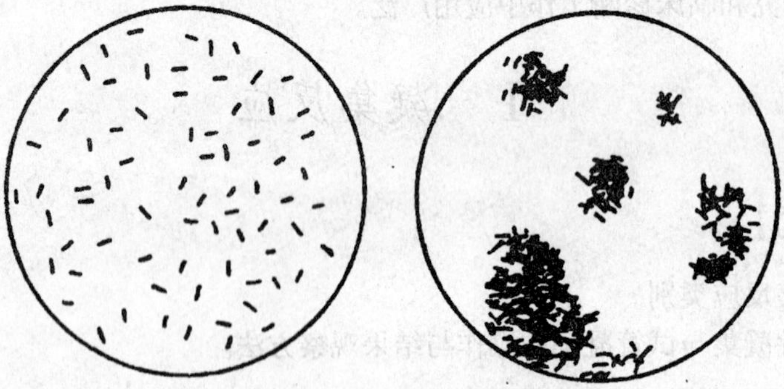

A　　　　　　　　B

图43－1　玻片凝集示意图

Fig. 43－1 The slide agglutination reaction

A 阴性反应；B 阳性反应

A. the negative results; B. the positive results

2. 试管凝集反应

（1）取洁净小试管8支，列于试管架上，依次注明号码，用1ml吸管吸取0.5ml生理盐水加于每支试管。

（2）用1ml吸管吸取伤寒杆菌“H”诊断血清0.5ml加入第1管中，用吸管反复吹吸3次，使血清与盐水充分混合，然后吸出0.5ml加入第2管，同样予以充分混合后吸出0.5ml加入第3管。依此类推，稀释到第7管。自第7管中吸出0.5ml弃去，第8管不加血清作对照，此时第1管到第7管的血清稀释倍数为1∶20，1∶40……至1∶1280，每支容量为0.5ml。

（3）用吸管吸取伤寒杆菌菌液，加入各管中，每管0.5ml（由第8管对照管开始向前加），晃动试管架混匀。此时血清稀释倍数又增加一倍，为1∶40，1∶80……至1∶2560。

（4）将试管置于37℃水浴箱水浴2～4小时，取出后观察试管底部结果，勿振摇。如所有的菌细胞都凝集成团沉于试管底部，可判断结果为4＋。对照管应混浊，如雾状。

（5）血清抗体效价为出现凝集现象的最大血清稀释倍数的倒数。

【结果与讨论】

1. 记录实验结果，图示玻片凝集反应。
2. 列表表示试管凝集反应的结果，并作出结论。
3. 如何利用凝集反应测定血型？

Ⅱ 沉淀反应

【实验目的】

1. 了解常见的沉淀反应的类别。
2. 掌握环状沉淀反应、单向和双向琼脂扩散实验的应用、操作与观察结果的方法。

【实验原理】

环状沉淀反应是在溶液中进行的简单的沉淀实验。将免疫血清加到试管底部，将稀释的含有可溶性抗原的材料小心重叠其上，让抗原与抗体在两液体的界面相遇，形成白色沉淀环。如抗原系列稀释后进行试验，可对抗原半定量测定。环状沉淀反应广泛用于法医学血迹鉴定和食品真伪鉴别等。

琼脂扩散法分单向和双向两种。单向琼脂扩散是指抗原或抗体这两种成分中只有一种成分扩散的方法。是目前最常用的一种抗原或抗体定量技术。将一定量的抗体混合于加热的琼脂中，倾注于玻板上，凝固后，在琼脂板上适当位置打孔，将抗原材料加入琼脂板的小孔内，孔中抗原向四周扩散，与琼脂板中的抗体相遇，在比例合适处呈现白色沉淀环，沉淀环的直径与抗原的浓度成正比。如先用不同浓度的标准抗原制成标准曲线，未知标本的抗原含量就可从标准曲线中求出。如将抗原加入琼脂而将抗体加入孔中，则可用来测定抗体的浓度。本法简便，灵敏度高，结果易于观察。

双向琼脂扩散指抗原和抗体在同一凝胶内扩散，彼此相遇后，在比例合适处出现特异性的沉淀线。双向琼脂扩散实验非但可对抗原或抗体进行定性鉴定和测定效价，还可对抗原或抗体进行纯度分析、分子量估计和同时对两种不同来源的抗原或抗体进行比较，分析其所含成分的异同等。

【实验材料】

1. 试剂

羊血清，抗羊血清，生理盐水，羊血清（已测浓度），3% 琼脂糖（以 pH 巴比妥 - 巴比妥钠缓冲液配），牛血清，马血清。

2. 其他

毛细吸管，沉淀反应用小试管，试管架，玻片，凝胶打孔器（直径 3mm），载玻片，水平台，水浴箱，37℃ 恒温箱，10ml 吸管，5μl 加样器。

【实验步骤】

1. 环状沉淀实验

（1）用细长的毛细滴管吸取抗体（抗羊血清），插入沉淀管管底，徐徐滴入，每管约0.2ml。

（2）用另一支毛细管吸取稀释的羊血清，轻轻重叠于抗羊血清上。注意务必使抗原和抗体间能显示出清楚的二层，不可产生气泡，以免抗原和抗体二者接触不均匀而影响结果。

（3）对照管以生理盐水（或人血清）代替羊血清。

（4）操作完毕后静置桌上，5分钟后，观察抗原和抗体的接触面，如有白色细微的环状沉淀，即为阳性。半小时后再观察一次。

2. 单向琼脂扩散实验

（1）取干净干燥载玻片，置水平台上备用。

（2）将3%琼脂糖融化，56℃水浴中保温。

（3）取1.5ml琼脂糖加到56℃预热玻璃管中，加入等量抗血清，充分混匀后铺于玻片上。

（4）待凝胶凝固后打孔（图43－2），孔距1.2～1.5cm，用记号笔在琼脂板的底面将孔编号。

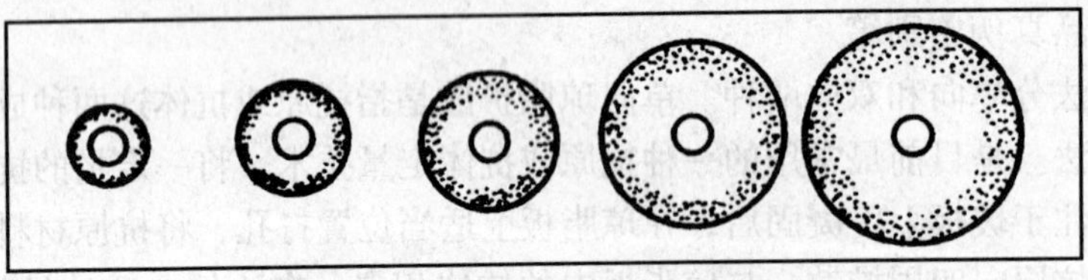

图43－2　单向琼脂扩散结果示意图

Fig. 43－2 The precipitin rings of simple immunodiffusion

（5）将系列稀释的标准羊血清和待检羊血清分别加到凝胶孔中，每孔5μl。

（6）将凝胶板置于湿盒内，37℃保温24小时后，观察结果，测量沉淀环直径。

（7）绘制标准曲线（图43－3），在其上标出待测羊血清浓度。

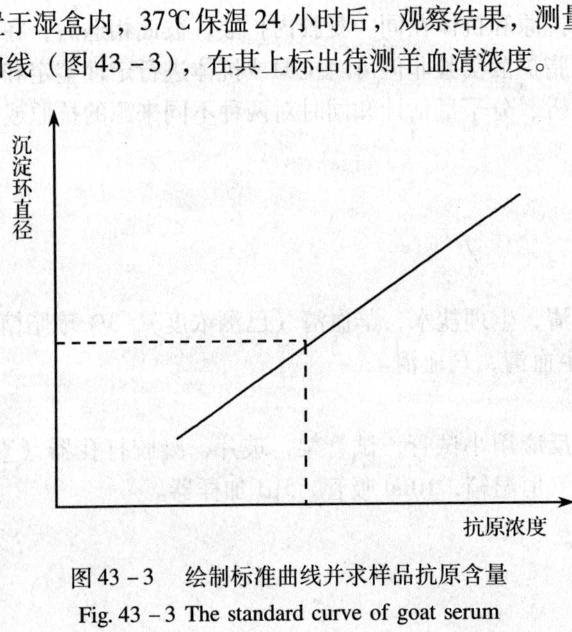

图43－3　绘制标准曲线并求样品抗原含量

Fig. 43－3 The standard curve of goat serum

3. 双向琼脂扩散实验

（1）取干净干燥平皿，置水平台上备用。

（2）将1%琼脂糖融化，取15ml琼脂糖铺于平皿上。

（3）凝胶凝固后打孔（图43－4），孔距1.0cm，用记号笔在琼脂板的底面将孔编号。

（4）中心孔中加5μl抗羊血清，周围分别加羊血清、牛血清、马血清和缓冲液各5μl（实验前可先将抗原、抗体做系列稀释，以获得适宜的抗原抗体比例）。

（5）将凝胶板置于湿盒内，37℃保温24小时后，观察结果。

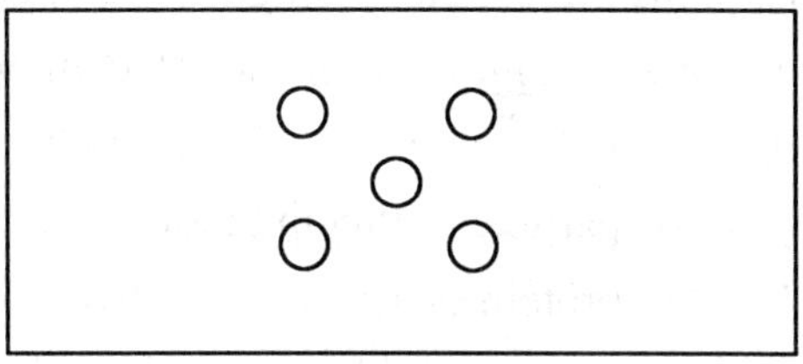

图43－4　双向琼脂扩散打孔图

Fig. 43－4 The holes of the double immunodiffusion

【结果与讨论】

1. 详细记录三个实验的结果。

2. 根据双向琼脂扩散实验的结果，分析羊血清与牛血清、马血清之间有无共同成分？抗羊血清的纯度如何？

43 Humoral Acquired Immune
——Agglutination Reaction, Precipitation Reaction

Antigen – antibody reactions are called serological techniques. They include agglutination and precipitation reaction. Agglutination is the combination of a particle antigen, such as bacteria, red blood cell, with specific antibodies, in the exist of electrolytes, will lead to a clumping of particles. Precipitation is the combination of a soluble antigen, such as polysaccharide, protein or lipoid, with specific antibodies, in the existence of electrolytes, will lead to an insoluble precipitate. Precipitation reaction is one of the basic methods in clinical tests. Agglutination reactions and precipitation reaction are classical serological reactions and are very useful in the research work and the diagnosis of certain diseases.

Ⅰ Agglutination Reactions

Purpose

1. To be familiar with the types of the agglutination reactions.
2. To perform the slide agglutination test and the tube agglutination test.

Principle

Agglutination reactions can be divided into direct agglutination, indirect passive agglutination, co – agglutination and anti – human globulin reaction. The direct agglutination reaction is the simplest agglutination and used widely, include the slide agglutination test and the tube agglutination test. The slide test is often used to diagnose antigens, such as unknown bacteria, with the known antiserum. It is the important method in detecting bacteria that lives in intestines, and also can be used in determining the human blood types. The most outstanding advantage of the slide test is rapid. The tube test is often employed to diagnose antibody with the known antigen. For example, the Widal tube agglutination test can identify *Salmonella*, and it is qualitative. In the tube test, a specific amount of antigen is added to a series of tubes, and serial dilutions of serum are then added to each tube and mixed. The greatest dilution of serum showing an agglutination reaction is determined, and the reciprocal of the dilution is the titer of the antibody in the serum.

Materials and Apparatus

1. Culture

Culture of *Salmonella typhimurium.*

2. Reagents

Salmonella typhimurium H antiserum (1/10 dilution), physiological saline.

3. Apparatus

Clean glass slide, inoculating loop, alcohol burner, wax pencil, clean glass slide, inoculating loop, alcohol burner, wax pencil, clean serological test tubes, 1ml pipette and test tube rack.

Procedures

1. The slide agglutination reaction

(1) Mark a line in the middle of the slide to divide it into two sections, mark one end of the slide S (*Salmonella*). Place a drop of the saline to each half of the slide.

(2) Place one loop of *Salmonella typhimurium* H antiserum on the S end, the other end adds saline as control.

(3) Suspend a loop of *Salmonella typhimurium* culture in the control end. Then suspend another loop of *Salmonella typhimurium* H antiserum in the S end. Mix each of the drops thoroughly.

(4) Pick up the slide and rock them back and forth gently. After a few minutes, observe the result with the unaided eye. (After observing, the slide should been sterilize.)

(5) If you see large clumps and the medium is clear, it happens as agglutination. The result is positive. The control result should be negative (Fig. 43 -1).

2. The tube agglutination reaction

(1) Using the wax pencil, label eight clean serology test tubes with the number 1 ~ 8. Pipette 0.5ml of physiological saline to each tube.

(2) Add 0.5ml of *Salmonella typhimurium* H antiserum to the first tube. Mix by drawing the solution into the serological pipette and slowly blowing it out for three times. Then transfer 0.5ml with the same pipette to the second tube. Mix an before and continue transferring 0.5ml aliquot to tube 3. Continue this procedure to tube 7, and discard 0.5ml from tube 7 after mixing. The tube 8 is the antigen control tube, without antiserum. Tubes 1 ~ 7 constitute a dilution series from 1/20, 1/40 through 1/1280. Each tube contains 0.5ml liquid.

(3) Add 0.5ml of *Salmonella typhimurium* culture to each tube (from tube 8 to tube 1). Mix by gently shaking the test tube rack to agitate all tubes. Tubes 1 ~ 7

constitute a dilution series from 1/40, 1/80 through 1/2560.

(4) Incubate the tubes in a 37℃ water bath for 2 ~ 4h. Then observe the result. Observe the bottom of each tube without shaking. A tube would be given a 4 + ranking if all cells had clumped and settled to the bottom of the tube. The control is seen as a compact bottom with a cloudy supernatant.

(5) The titer of the antiserum is the reciprocal of the highest dilution showing agglutination.

Results and Discussion

1. Draw the result of the slide agglutination reaction.
2. Tabulate the tube agglutination reaction result.
3. How can agglutination reactions be used to identify blood type?

Ⅱ Precipitation Reactions

Purpose

1. To be familiar with the precipitation reactions.
2. To perform the ring test, the simple diffusion test and the double diffusion test.

Principles

The ring test is a simple precipitation reaction that illustrates in solution. Antiserum is introduced into a small - diameter test tube, and the antigen is then carefully added to form a distinct layer at the interface between the two reagents. The antigen can be semi - quantitative, if a series of dilutions of the antigen is used. This method is extensively used in the bloodstain identification in forensic medicine and foodstuff.

Immunodiffusion includes simple diffusion test and double diffusion test. In the simple diffusion test, either antigen or antibody mixed through the support of diffusion. It can be used for quantitative analyze antigen or antibody. Mix some antibody with thawed agar and pour them in slides. After solidify, make wells on agar and place antigen in wells. The antigen in wells can diffuse. The precipitating ring will be found. The direct proportion can be found between the diameter of precipitating rings and the antigen. Draw a standard curve with an antigen with known concentration and the diameter of precipitating rings and then the unknown concentration of antigen can be obtained from the curve. If mix the agar with antigen and add the antibody in wells, it can be used to determine the concentration of antibody. It's a simple, high - sensitivity method. And the results are easy to be observed.

In the double diffusion, antigen and antibody are placed and diffused in the same agar. As they meet, the precipitation lines will be found in the area with suitable concentrations. The double diffusion can be used to qualitative analyze the antigen or antibody, determine the titres, analyze purity, estimate molecular weigh, compare the similarities and differences of antigen or antibody from different sources.

Materials and Apparatus

1. Reagent

Goat serum, anti - goat whole serum, 0.85% saline, standard goat serum (with known concentration), goat serum (with unknown concentration), 3% agarose (make up with pH8.6, 0.1mol/L barbital - barbital sodium buffer), bovine serum, horse's serum.

2. Apparatus

Capillary, tube, test tube rack, slides, cork borer (with the diameter of 3mm), horizontal table, Petri dishes, water bath, 37℃ incubator, pipettes.

Procedures

1. The ring test

(1) Add 0.2ml antigoat serum with capillary to tubes.

(2) Add goat serum with other capillary carefully to form a layer over the antiserum. Avoid bubbles.

(3) The control tube replaces goat serum with saline or human serum.

(4) After 5min, observe the interface between the serum and antiserum. There positive tubes have a white precipitating ring. After 30 min, observe again.

2. The simple diffusion

(1) Put clear and dry slide on horizontal table.

(2) Thaw 3% agarose and keep warm in 56℃ water bath。

(3) Add 1.5ml agarose to 56℃ preheat tubes, then add 1.5ml antiserum. Mix it and pour a slide.

(4) After the agar has hardened, cut holes with a cork borer (Fig. 43 - 2) with the center of the holes about 1.2 ~ 1.5cm apart, label the slides.

(5) Add 5μl of a set of serial diluted standard goat serum and the goat serum (with known concentration) into the holes.

(6) Incubate the slide at 37℃ for 24 hours in a damp Petri dish. Then observe results, measure the diameter of precipitating rings and calculate the concentration of the antibody in the goat serum (Fig. 43 - 3).

3. The double immunodiffusion

(1) Put clear and dry slide on horizontal table.

(2) Thaw 1% agarose and pour 15ml agarose to a Petri dish.

(3) After the agar has hardened, cut holes with a cork borer (Fig. 43 – 4) with the center of the holes about 1.0 cm apart, label the holes.

(4) Add 5μl antigoat serum in the central hole. Add 5μl goat serum, bovine serum, horse serum and buffer in surrounding holes.

(5) Incubate the dishes at 37℃ for 24h in a damp box. Then observe results.

Results and Discussion

1. Record results of these three experiments.

2. According to the results of the double diffusion, analyze whether there is common ingredients in goat serum, bovine serum and horse serum and purity of the antigoat serum.

实验四十四　体液免疫——免疫标记技术

免疫标记技术是将抗原、抗体的特异性结合与放射性同位素、酶、荧光、胶体金等的标记结合在一起的方法。该项技术可定性、定量、定位检测抗原或抗体。具有特异、灵敏、快速、结果易于观察等特点，已越来越广泛地应用于医学和其他生物学科的研究领域及临床检验工作。目前有许多用相关的商品试剂盒供应。本试验介绍了酶联免疫吸附试验及胶体金标记技术的原理和试验方法。

Ⅰ　间接 ELISA 法检测 AFP

【实验目的】

掌握 ELISA 的原理和实验方法。

【实验原理】

酶联免疫吸附试验是以酶标记的抗原或抗体去检测液体中的未知抗体或抗原，随后加入酶的底物，使产生有色产物，颜色的深浅与抗原或抗体的量成正比，可直接目测结果，也可通过酶标仪进行定性定量测定。此法可定性，也可定量，在实验室研究中应用非常广泛。标记酶及底物一般可直接购买。酶联免疫法有间接法、双抗体法等。本试验利用双抗体法检测样品中的 AFP（甲胎蛋白）。

【实验材料】

1. 试剂

鼠抗 AFP 单抗，羊抗鼠 HRP 酶标 IgG，酶作用底物（A 液：联苯二胺；B 液：过氧化氢），洗涤液（pH7.4 0.01mol/L Tris - HCl 缓冲液），终止液（2mol/LH_2SO_4），AFP 标准品，阳性对照血清（来自孕妇），阴性对照血清（来自健康人群），血清样品。

2. 其他

已包被抗 AFP 抗体的聚苯乙烯酶联免疫反应用小皿，移液管。

【实验方法】

1. 向反应用小皿中加入血清或 AFP 标准品 2 滴，室温下 3 分钟后弃去，洗涤液洗 3 次，弃去。

2. 加入 1 滴酶标抗 AFP，室温下 5 分钟后弃去，水洗 3 遍，弃去。

3. 加入 1 滴 A 液和 1 滴 B 液，5 分钟后加入终止液 1 滴，观察结果。黄色者为阳性。

【结果与讨论】

1. 记录本次实验结果。
2. 为什么可在妊娠女性血液中检测到 AFP？AFP 检测的意义何在？

Ⅱ 胶体金技术检测 hCG

【实验目的】

掌握胶体金技术快速检测抗原的方法。

【实验原理】

人类绒毛膜性腺激素（hCG）是受精后，胎盘发育过程中分泌的一类糖蛋白激素。由于 hCG 在怀孕后立刻出现且浓度在孕早期快速升高，使得它成为孕早期检测的极佳标记。胶体金技术是一个简便快速的抗原检测方式，可用于检测尿液样本中的 hCG 含量。妊娠检测试纸条是胶体金技术应用于早早孕检测的商品化试剂盒。

【实验材料】

尿液收集杯，早早孕试纸条。

【实验方法】

1. 以小杯收集尿液样本。
2. 将试纸条垂直小心地插入样本液中。将标有箭头的一端对准尿液。注意不要将 MAX 线（标记线）位置浸没。
3. 约 3 秒钟后取出试纸条，平摊于洁净、干燥、不吸水的表面。
4. 待反应完全结束，着色带显现。读取结果（参见图 44－1）。

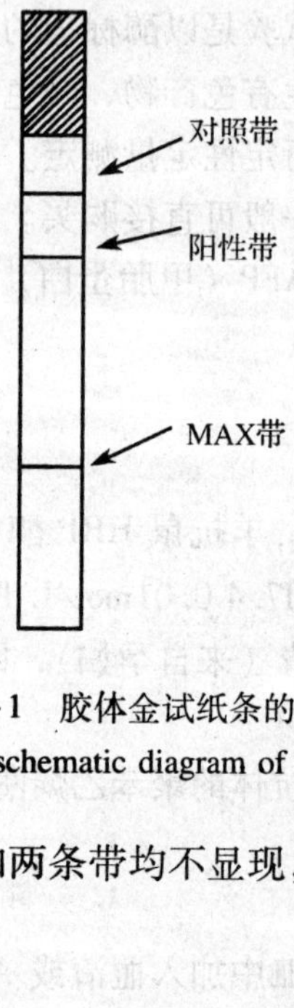

图 44－1　胶体金试纸条的结果示意图。

Fig. 44－1 The schematic diagram of the colloidal gold strip

不同样品的检测可能导致 3 种不同的结果产生。如试纸条上出现对照带和阳性带，表明样品为 hCG 阳性。如试纸条上仅出现对照带，表明样品为 hCG 阴性。如两条带均不显现，表明试纸条失效。

【结果与讨论】

1. 记录本次实验结果。
2. 为什么 hCG 可用于妊娠期的检测？

44 Humoral Immune
——Immunolabelling Technique

Immune label method is a combination of the specific reaction of antigen and antibody and the marking methods with isotope, enzyme, fluorescence or colloidal gold. This method can be used to qualitative analysis, quantitative analysis or position analysis of antigen and antibody. Immunelabelling technique is a more specific, sensitive, rapid method and easier to been observed. It was more and more widely used in medicine and other fields of biological study and clinical test. There are a number of related commercialization kits appear on the market. In this exercise, enzyme - linked immunoabsorbent assay (ELISA) and Colloidal gold technique are introduced.

I Indirect ELISA of detecting AFP

Purpose

To demonstrate a method for the identification of antigens or antibodies by using ELISA.

Principle

The procedure of ELISA is predicated by the use of an enzyme - linked specific antigen or antibody to react with unknown antibody or antigen, then add the substratum of the linked enzyme. The products of enzyme's color can be observed by eye or spectrophotometrically. This method is a widely accepted method for the detection of specific antigens or antibodies. And it can be used to qualitative analysis or quantitative analysis. The enzyme, such as HRP, and the substratum, such as OPD and H_2O_2, can be obtained easily. This test can be performed as an indirect immunoabsorbent assay, a double - antibody technique *etc.* In this exercise, it will be used to detect AFP in samples.

Materials and Apparatus

1. Reagent

Mouse anti - AFP McAb, goat anti mouse IgG/HRP, substratum (A - OPD and B - H_2O_2), washing buffer (pH7.4, 0.01mol/L Tris - HCl), 2mol/L H_2SO_4, standard AFP, the positive serum (from pregnant woman with 1 ~4 months pregnancy), the negative control serum (from healthy man), serum samples.

2. Apparatus

Microtiter plates (binding anti – AFP antibody), clear pipettes.

Procedures

1. Drip 2 drops of serum or standard AFP into microtiter plates that bind anti – HCG antibody. Incubate in room temperature for 3 min, then discard it. Wash with washing buffer for three times.

2. Drip a drop of mouse anti – AFP McAb that labeled with HPR. Incubate in room temperature for 5 min, then discard it. Wash for three times.

3. Add a drop of OPD and a drop of H_2O_2. After 5min, add a drop of 2 mol/L H_2SO_4 to stop the reaction. Observe results, the positive result is yellow.

Results and Discussion

1. Record results of this exercise.

2. Why the AFP can be measured in pregnant woman? What does the AFP detection means?

II Colloidal gold technique of detecting hCG

Purpose

To demonstrate a method for the identification of antigens by using colloidal gold technique.

Principle

Human chorionic gonadotropin (hCG) is a glycoprotein hormone secreted by the developing placenta shortly after fertilization. The appearance of hCG soon after conception and its subsequent rise in concentration during early gestational growth make it an excellent marker for the early detection of pregnancy. Colloidal gold technique is a simple and rapid detection of antigen. It can be used to test the hCG in Urine specimen. Pregnancy Test Strip is a test kit for the determination of hCG to diagnosis the early pregnancy.

Materials and Apparatus

Sample collection cups, hCG Strips.

Procedures

1. Collect the urine specimens for the experiment in a sample collection cup.

2. Holding the strip vertically, carefully dip it into the specimen. Immerse the

strip into the urine sample with the arrow end pointing towards the urine. Do not immerse the MAX Line (Marker Line) .

3. Take the strip out after 3 seconds and lay the strip flat on a clean, dry, non-absorbent surface.

4. Wait for colored bands to appear until the reaction completely finish. Then read results (see Fig. 44 – 1) .

There are three results can be read with different samples. If the strip appear control line and positive line, it means the sample is hCG positive. If the strip appear only control line, the sample is hCG negative. If none of these two lines are appear, it means the strip is failure.

Results and Discussion

1. Record results of this experiment.
2. Why the hCG test can be used to judge pregnancy?

实验四十五　细胞免疫检测——体外试验

细胞免疫担负着迟发型变态反应、移植排斥、肿瘤免疫等。细胞免疫功能主要由T细胞完成。检测T细胞的数量、功能及其产物，是了解机体细胞免疫状态的重要手段。本实验介绍了可在体外用于鉴别T细胞数量、检测T细胞功能的方法。

Ⅰ　E玫瑰花环实验

【实验目的】

1. 熟悉鉴别T淋巴细胞的方法。
2. 掌握E玫瑰花环实验方法。

【基本原理】

T淋巴细胞鉴定有多种方法。人的T细胞可以结合绵羊红细胞，形成玫瑰花环。E玫瑰花环实验不但可用来检测人T细胞的免疫功能，还是一种重要的T细胞计数方法。B淋巴细胞可通过补体参与的EAC玫瑰花环实验进行鉴定。

【实验材料】

1. 细胞

外周血单个核细胞（PBMC，以RPMI1640悬浮）。

2. 试剂

绵羊红细胞（SRBC，用阿氏液保存），Hanks液，RPMI1640培养液，胎牛血清（FCS，56℃灭活1小时），无菌双蒸水，吉姆萨-瑞氏染液。

3. 其他

离心机，洁净载玻片，吸管，圆底离心管，显微镜。

【实验方法】

1. 取10ml SRBC，置于20ml离心管中。用Hanks液装满离心管，1000×g离心10分钟。去上清，加Hanks液，以吸管悬浮细胞。1000×g离心10分钟。重复上述洗涤过程1次。

2. 以RPMI1640悬浮SRBC至终浓度2%（体积分数）。4℃保存（可保存5~7日）。

3. 将$\leq 2\times10^7$ PBMC置圆底离心管中，400×g离心10分钟。去上清，以RPMI1640悬浮细胞至终浓度1×10^7 cells/ml。

4. 将2ml FCS，2ml SRBC，1×10^7 PBMC混匀。混合物在37℃水浴孵育10分钟。

200 × g 4℃离心 5 分钟。在冰上孵育 1 小时。

5. 悬浮沉淀，加 0.8% 戊二醛 0.1ml，混匀，4℃固定 15 分钟。

6. 在洁净载玻片上滴一小滴细胞悬液。空气干燥后，以吉姆萨－瑞氏染液染色。冲洗、干燥后，显微镜观察 PBMC 中的 T 淋巴细胞。

【结果与讨论】

1. 图示 E 玫瑰花环。
2. 简述其他计数 T 淋巴细胞的方法。

Ⅱ 淋巴细胞体外转化试验

【实验目的】

1. 熟悉淋巴细胞转化试验的原理。
2. 掌握 T 淋巴细胞体外转化实验的方法。

【实验原理】

当机体被抗原或有丝分裂原（如结核菌素或 PHA 等）刺激后，淋巴细胞能够转化为淋巴母细胞，并进一步发育成致敏淋巴细胞。在一定程度上，这种转化现象代表着机体的细胞免疫状态。通过加入合适的刺激物，T 细胞在体外发生转化并显示出形态上的改变。目前，常用的检测细胞免疫功能的方法是 PHA 诱导的 T 淋巴细胞转化试验。可通过三种不同的方法检测细胞转化率：形态学检查、MTT 法和同位素标记法。本实验采用的是形态观察法。

【实验材料】

1. 细胞

外周血单个核细胞（PBMC，以 RPMI1640 悬浮）。

2. 试剂

吉姆萨－瑞氏染液，RPMI1640 培养液，肝素（400U/ml），PHA（1mg/ml），2.5% 碘酒，75% 乙醇。

3. 其他

玻片，无菌棉签，无菌注射器，毛细吸管，96 孔培养板，CO_2 恒温培养箱，移液管。

【实验方法】

1. 以 RPMI1640 悬浮 PBMC 至终浓度 1×10^6 cells/ml。在 96 孔板的每孔中加入 100μl。

2. 在孔中加入 PHA 至终浓度 10μg/ml。每个样品制 3 个平行孔。

3. 37℃孵育 72 小时。

4. 离心 3000 转/分钟，10 分钟，去上清。

5. 以残余液体悬浮细胞，取少量滴于玻片上，涂片，干燥。

6. 吉姆萨－瑞氏染液染色。

7. 显微镜观察淋巴细胞的转化情况，一般计数 100～200 个淋巴细胞，然后计算转化率。正常情况下，PHA 刺激淋巴细胞的转化率在 60%～80%。

（转化的淋巴细胞形态特点：体积增大，比原来的成熟淋巴细胞大 2～3 倍；核膜清晰，染色体疏散呈网络状结构；胞浆丰富，染色为嗜碱性。）

【结果与讨论】

1. 计算淋巴细胞的转化率。
2. 在何种条件下，T、B 淋巴细胞会发生转化？

45 Cellular Immune Test——*in vitro*

Cellular immunity responsible for the delayed type hypersensitivity, allograft rejection, tumor immunity. T cell is the main responsible of cellular immune function. Detection of the number, functions, products of T cells can help us to realize the cellular immune status. In this experiment, how to identify the number of T cells and how to detect the function of T cell *in vitro* are introduced.

Ⅰ Erythrocyte rosette Test

Purpose

1. To be familiar with the methods about indification of T cell.
2. To perform the erythrocyte rosette formation test.

Principle

There are many methods about indification of human T cell count. The erythrocyte rosette formation test can be used to detect the immune function in human and animal cells. And it's also an important method in T cell count, as T cell can bind to and form rosettes with sheep red blood cells (SRBC). An EAC rosette formation test is used to identify B cell.

Materials and Apparatus

1. Culture

Peripheral blood mononuclear cells (PBMC in complete RPMI1640 medium).

2. Reagents

SRBC in Alsevers solution, Hanks (balanced salt solution), complete RPMI1640 medium, fetal calf serum (FCS, heat-inactivated 1h at 56℃), Sterile distilled water, Giemsa-Wright dye.

3. Apparatus

Centrifuge, clean glass slide, clean glass slide, pipette, round-bottom centrifuge tube, microscope.

Procedures

1. Place 10ml SRBC in Alsevers solution in a centrifuge tube. Fill tube with

Hanks and centrifuge 10 min at 1000 × g. Remove supernatant, and using a pipet, resuspend cells in Hanks . Centrifuge 10 min at 1000 × g. Repeat this washing once.

2. Add complete RPMI1640 to the pelleted SRBC (final concentration 2% *V/V*) . Store at 4℃ until ready to use (can be stored for 5 to 7 days) .

3. Transfer $\leqslant 2 \times 10^7$ PBMC to a round – bottom centrifuge tube. Centrifuge 10 min at 400 × g) . Remove supernatant and resuspend cells in complete RPMI1640 for a final concentration of 1×10^7 cells/ml.

4. Add 2ml heat – inactivated FCS and 2 ml of SRBC for each 1×10^7 PBMC. Incubate the mixture 10 min in a 37℃ water bath. Centrifuge 5 min at 200 × g, 4℃. Incubate 1h on ice.

5. Resuspend the precipitation and add 0. 8% glutaraldehyde 0. 1ml, mix them and fix at 4℃ for 15min.

6. Drop a small drop of suspension on a clean slide. After air dry, stain it with - Giemsa – Wright dye. Wash, dry and then observe with microscope.

Results and Discussion

1. Draw the erythrocyte rosettes.

2. Are there any other methods can used to count T cell?

Ⅱ Lymphocyte transformation test *in vitro*

Purpose

1. To be familiar with lymphocyte transformation principle.
2. To perform the lymphocyte transformation test *in vitro*.

Principles

When the body is stimulated by antigen or mitogen (such as the tuberculin or PHA, *etc.*), lymphocytes can be transformed into lymphoblastoid cells, and further into sensitized lymphocytes. To a certain extent, this transformation represents the state of the body's cellular immune. By adding appropriate stimulus, T cell also can be transformed *in vitro* and showed morphological changes. At present, the regular method for examining the cellular immune of is the PHA – induced T lymphocyte transformation test. There are three specific ways: morphological examination, MTT assay and isotopic tracer method. The last method is the most accurate and repeatable but more complex and expensive. In this experiment, the morphological is been observed.

Materials and Apparatus

1. Culture

Peripheral blood mononuclear cells (PBMC in complete RPMI1640 medium).

2. Reagent

Giemsa – Wright dye, RPMI1640, heparin (400unit/ml), PHA (1mg/ml), 2.5% iodine, 75% alcohol.

3. Apparatus

Slides, sterile cotton, sterile syringe, capillary, 96 – well plate, CO_2 incubator, pipettes.

Procedures

1. Resuspend PBMC in complete RPMI1640 to a final concentration of 1×10^6 cells/ml. Add 100μl to each hole of 96 – well plate.
2. Add PHA 100μl, final concentration 10μg/ml, to each hole. Three holes used for each sample.
3. Incubate at 37℃ for 72h.
4. Centrifuge 3 000r/min for 10min, remove supernatant.
5. Resuspend cells with remnant liquid. Drip a drop on slide, smear and dry.
6. Stain with Giemsa – Wright dye.
7. Observing with microscope, counting at least 100 ~ 200 lymphocytes, then calculating the conversion rate. Normal conversion rate is at 60% ~ 80%.

Morphological characteristics of transformated lymphocyte: volume increased about 2 to 3 times than the original maturity lymphocytes, nuclear membrane clearly visible and nuclear chromatin was loose as network structure, cytoplasm rich and basophilic staining.

Results and Discussion

1. Calculate the lymphocyte conversion rate.
2. In which circumstances, T, B lymphocytes can occur transformation?

实验四十六　细胞免疫——体内实验

用于测定细胞免疫功能的方法有体内法和体外法。体内法可采用体内淋巴细胞转化、迟发型变态反应等检测机体细胞免疫功能。迟发型变态反应的方法应用较广，可以正确反映机体细胞免疫状况。人体的迟发型变态反应的检测常用皮试法，动物多采用耳廓肿胀法等等。

Ⅰ　豚鼠结核菌素试验

【实验目的】

掌握结核菌素试验的方法。

【基本原理】

结核菌素是结核杆菌的组成成分。人体感染结核杆菌或接种卡介苗后，将产生致敏T细胞。致敏T细胞能够释放大量淋巴因子，并形成单核细胞浸润的炎症反应。

豚鼠对结核杆菌具有高度敏感性，其感染后的病理变化与人类疾病相似。豚鼠结核菌素试验是分离、鉴定和诊断结核杆菌，筛选抗结核杆菌药物的最佳动物模型。

【实验材料】

豚鼠两只，1∶1000稀释的旧结核菌素（OT），卡介苗注射器及针头，剪刀，碘酒棉球，酒精棉球。

【实验方法】

1. 试验正式进行前两个月，准备两只豚鼠。其中一只皮内注射卡介苗致敏。
2. 以卡介苗注射器对两只豚鼠分别注射0.1ml 1∶1000的旧结核菌素。
3. 注射后48～72小时，观察注射位置的皮肤红肿、肿块大小等反应。

【结果与讨论】

1. 记录结核菌素试验的结果。
2. 简述结核菌素试验的技术和临床意义。

Ⅱ　体内淋巴细胞转化试验

【实验目的】

掌握体内淋巴细胞实验的方法。

【基本原理】

外周血 T 淋巴细胞在体内遇到 PHA 或 ConA，将转化为淋巴母细胞。通过外周血涂片染色，可进行淋巴细胞计数和转化率计算。转化率将反映机体细胞免疫水平，可作为机体免疫功能检测的指标。

【实验材料】

1. 试剂

吉姆萨－瑞氏染液，ConA （0. 5mg/ml ），heparin（400unit/ml）。

2. 其他

玻片，棉球，离心机，显微镜，细胞计数板。

【实验方法】

1. 试验正式进行前 3 天，对实验组小鼠腹腔注射 ConA 0. 3 ~0. 5 mg。
2. 摘取小鼠眼球，取外周血，滴加入预先加入肝素的试管。
3. 涂片：取一滴外周血滴入玻片中央。推片，干燥。
4. 固定：滴加 1 ~2 滴甲醇，干燥。
5. 染色：滴加 2 滴吉姆萨－瑞氏染液，同时滴加两滴水。染色 5 ~10 分钟。
6. 洗涤，然后用吸水纸轻轻吸干玻片上的水分。
7. 显微镜观察，计数转化的细胞，然后计算转化率。

$$转化率=\frac{转化的淋巴细胞数量}{转化的淋巴细胞数量+未转化的淋巴细胞数量}\times 100\%$$

【结果与讨论】

计算并比较试验组和对照组的淋巴细胞转化率。

46 Cellular Immune ——*in vivo*

The cellular immune function can be measured *in vivo* and *in vitro*. Commonly used *in vivo* methods are lymphocyte transformation, delayed hypersensitivity testing method *etc.* Application of these methods are used wider, can accurately reflect the state of cellular immunity. Human commonly used testing of delayed type hypersensitivity are skin test. The animals are ear swelling method.

Ⅰ Tuberculin test in guinea pigs

Purpose

To master Tuberculin test method.

Principle

Tuberculin is components of Mycobacterium tuberculosis. If the body have been infected by TB infection or vaccinated by BCG, sensitized T cell will produced. The tuberculin sensitizated T cells can release a variety of lymphokines and will form monocyte – based inflammatory cell infiltrationin.

Guinea pigs have a high degree of sensitivity of *Mycobacterium tuberculosis.* And its pathological changes after infection are similar to human disease. *Mycobacterium tuberculosis* in guinea pigs is the best animal model for separation, identification, diagnosis tubercle bacillus and screening of anti – TB drugs.

Materials and Apparatus

Two guinea pig, 1 : 1000 dilution of old tuberculin (Old tuberculin, OT), BCG syringes and needles, scissors, iodine swab, alcohol swab

Procedures

1. Prepare two guinea pigs for two months. One of them is sensitized by an intradermal injection of BCG.

2. Use BCG syringe to inject 0.1ml of 1 : 1000 old tuberculin in two of the guinea pig respectively.

3. After injection for 48 ~ 72h, observe the skin reactions in the injection site, such as whether there is mass, and how big.

Results and Discussion

1. Record the result of of tuberculin test.

2. Briefly describe the technique and clinical significance of tuberculin test.

Ⅱ Lymphocyte transformation test *in vivo*

Purpose

To perform the lymphocyte transformation test *in vivo*.

Principles

When encounter to PHA or ConA, the peripheral blood T lymphocytes *in vivo* will transform to lymphoblastoid cell. Through the peripheral blood smear staining, lymphocyte count the conversion rate can be calculated. Conversion rate may reflect the level of cellular immunity, and often seen as indicators of cellular immune function.

Materials and Apparatus

1. Reagent

Giemsa – Wright dye, ConA (0.5mg/ml), heparin (400unit/ml).

2. Apparatus

Slides, straw, cetrifuge, microscope, cell count board.

Procedures

1. Three days before the experiment, the mouse of experimental group was intraperitoneal injection ConA 0.3 ~ 0.5mg.

2. During the experiment, collection of peripheral blood of mice through remove of the eyeball, and then add it to the heparin pre – addition test tube.

3. Smear: Take a small drop of blood in the central of slide. Push open the blood. Air dry.

4. Fix: drip 1 ~ 2 drops of methanol on the smear, air dry.

5. Stain: add 2 drops of Giemsa – Wright dye on the smear, plus 2 drops of water at the same time. Dye for 5 ~ 10 min.

6. Wash and then gently dry the slide with absorbent paper.

7. Observate transformed lymphoid cell with microscope. Count the transformed cell and then calculate the conversion rate.

$$\text{Conversion rate} = \frac{\text{Number of transformed lymphocytes}}{\text{Number of transformed lymphocytes} + \text{Number of non-transformed lymphocytes}} \times 100\%$$

Results and Discussion

Calculate and compare lymphocyte conversion rate of the experimental group and control group mice.

实验四十七　细胞因子检测

细胞因子是相关细胞分泌的具有生物活性的小分子蛋白质。免疫应答过程中，细胞因子在许多生理和病理过程中发挥重要作用，如免疫调节、炎症反应、肿瘤转移等。细胞因子的检测不仅用于基础免疫学的研究，而且还为临床诊断、疗效观察等提供参考指标。但由于含量低，细胞因子的诊断具有一定难度。检测细胞因子的常用方法有三种，包括生物活性检测，免疫学和分子生物学检测。

【实验目的】

1. 熟悉生物活性法检测 IL－2 的原理。
2. 掌握生物活性法 IL－2 检测的技术及计算方法。

【基本原理】

白细胞介素 2（IL－2）是一种 T 细胞源性的细胞因子，对多种免疫细胞都有促进生长和促进分化的作用。一些肿瘤细胞株必须依赖于 IL－2 才能生长，如 CTLL、CTB6、F12、CTLL－2 等。因此本实验以 CTLL－2 的增殖情况来检验IL－2 的生物活性。

MTT（四甲基偶氮唑蓝）法是实验室常用的检测手段，通过标准比色法可以测量琥珀酸脱氢酶活性（该酶可将 MTT 降解为紫蓝色的结晶）。琥珀酸脱氢酶的活性不但与细胞生长和活力成正比，而且可用于检测潜在药物和其他有毒物质对细胞的毒性。

【实验材料】

1. 细胞

CTLL－2 细胞。

2. 试剂

IL－2 样品，IL－2 标准品，5mg/ml MTT 溶液，10% 牛血清 RPMI1640 培养液，DMSO。

3. 其他

离心机，离心管，洁净载玻片，移液管，显微镜，毛细吸管，96 孔板，细胞计数板，CO_2 培养箱，酶标仪（570 nm 和 630 nm 滤光片）。

【实验方法】

1. 用 10% 牛血清 RPMI－1640 培养液洗涤 CTLL－2 细胞 2 次，1000 转/分，离心 10 分钟，移去含 IL－2 的原始培养基质。
2. 以含 10% 牛血清的 RPMI1640 调整细胞浓度至 1×10^5/ml。
3. 将 IL－2 样品和标准品不同稀释度加入 96 孔板中。每孔 50μl，设置 3 个复孔。同时设培养基质对照。

4. 加 CTLL－2 细胞悬浮液 50μl/孔。37℃ 5% CO_2 培养 24 小时。

5. 加 10μl MTT 溶液至每孔，继续孵育 4 小时。

6. 移去 50μl 上清，然后添加 50μl DMSO，置室温 10～20 分钟。混匀使产物完全溶解。

7. 1 小时后，以酶标仪测 OD 值。取 3 个复孔的均值为每种样品的 OD 值。最终 OD 值为 OD570nm～OD630nm，再减去培养液对照孔 OD 值。

$$\text{IL}-2\text{样品活性单位}(\text{U/ml})=\frac{\text{标准品最大 OD 值 50\% 的样品稀释度}}{\text{标准品最大 OD 值 50\% 的标准品稀释度}}\times\text{IL}-2\text{标准品活性}$$

【结果与讨论】

1. 计算 IL－2 样品的活性单位。
2. 常用检测细胞因子活性的不同方法有哪些？

47 Detection of Cytokines Levels

Cytokines are small molecule proteins secreted by cells with biological activity. In the course of immune responses, cytokines play an important role in many physiological and pathological processes, such as immune regulation, inflammatory response, tumor metastasis, *etc.* Detection of cytokines is not only for the basic immune research, but also for clinical diagnosis, the treatment observe, *etc.* But because of low content, cytokins are difficult to detect. There are three methods usually used to detect cytokins. Those are bioactivity detection, immunological detection and molecular biology methods.

Purpose

1. To be familiar with the principle of bioactivity detection of IL – 2 .

2. To perform the bioactivity detection and calculation of IL – 2.

Principle

Interleukin – 2 (IL – 2) is a T cell – derived cytokine known to promote growth and differentiation of many immune cells. The growth of some tumor cell lines are depended on IL – 2, such as CTLL, CTB6, F12, CTLL – 2. So the biological activities of IL – 2 can been tested by the CTLL proliferation.

The MTT assay is a laboratory test and a standard colorimetric assay for measuring the activity of succinate dehydrogenase that reduce MTT to formazan. The activity of succinate dehydrogenase have direct proportion to cell proliferation. It can be used to determine cytotoxicityof potential medicinal agents and other toxic materials.

Materials and Apparatus

1. Culture

CTLL – 2 cells.

2. Reagents

IL – 2 sample, IL – 2 standards, 5mg/ml MTT solution, complete 10% FCS – RPMI1640 medium, acidified isopropanol.

3. Apparatus

96 – well plate, pipette, capillary, centrifuge, centrifuge tube, clean glass slide,

microscope, cell count board, CO_2 incubator, microtiter plate reader with 570 – nm filter and optionally 630 – nm filter.

Procedures

1. Wash CTLL – 2 cells with 10% FCS – RPMI – 1640 culture medium twice. centrifuge at 1000r/min for 10min to remove the original growth medium, which contains IL – 2.

2. Adjuste cell concentration to 1×10^5/ml with 10% FCS – RPMI – 1640 culture medium.

3. Add the dilutions of the IL – 2 sample and the standard respectively in 96 – well plate, 50μl per well in triplicate. At the same time, set the culture medium control.

4. Add 50μl/hole with CTLL – 2 cell suspension, cultivate at 37℃ and 5% CO_2 for 24h.

5. Add 10μl MTT solution to each hole and continue to foster 4h.

6. Discard 50μl supernatant, then add 50μl DMSO and placing at room temperature 10 ~ 20min. Mix it to make the product fully dissolved.

7. After 1hour, measure OD values of with microtiter plate reader. OD value of each sample shall be taken as the average of 3 parallel hole. The final OD value = OD570nm – OD630nm – the contro OD.

The activity units of IL – 2 samples (U / ml)

$$= \frac{\text{the sample dilution corresponding to the 50\% maximum OD value of standard}}{\text{the standard dilution corresponding to the 50\% maximum OD value of standard}} \times \text{the activity of IL} - 2 \text{ standard}$$

Results and Discussion

1. Calculate the activity units of IL – 2 samples .
2. To understand different methods to detect the acticity of cytokines?

实验四十八　免疫调节药物的研究

免疫调节药物的药效研究一般采用体外和体内相结合的方式。体外试验研究可明确药物对免疫应答某一特定环节的具体影响，而体内研究则可探讨药物对机体免疫应答水平，对正常免疫应答、异常免疫应答等的综合影响。研究药物对机体免疫系统和免疫功能的作用，不仅能够深入揭示药物的作用机制，更为疾病的药物治疗提供基础。

免疫调节药物的初步研究主要包含四部分内容：建立动物模型；药物对固有性免疫功能的影响；药物对获得性体液免疫功能的影响和药物对获得性细胞免疫功能的影响。

实验动物的选取，除正常动物外，往往选取一种以上的免疫抑制（多采用环磷酰胺、氢化可的松等抑制剂）或功能低下（通过辐射等方法可获得）的动物模型，以评价药物的作用。

药物对固有性免疫功能的影响多以吞噬细胞功能、溶菌酶功能等为研究对象。具体实验方法可参考实验四十一、四十二。

药物对获得性体液免疫功能的影响可从多个方面进行研究。采用单扩、散射比浊等方法，可对血清中 IgG、IgA、IgM 等类别进行测定；以血清溶血素测定法、ELISA 等方法能够检测各种特异性抗体；溶血空斑试验、溶血分光光度法等试验可用于检测抗体产生细胞。本实验介绍了溶血空斑试验。

药物对获得性细胞免疫功能的影响主要通过淋巴细胞转化实验、接触性过敏反应来检测。本实验介绍了以 DNFB 诱发的接触性过敏反应。

Ⅰ 溶血空斑试验

【实验目的】

1. 熟悉溶血空斑试验的原理。
2. 掌握溶血空斑试验的方法。

【实验原理】

溶血空斑实验是体外检测单个 B 淋巴细胞（抗体形成细胞）的一种方法。经绵羊红细胞（SRBC）免疫后的 B 淋巴细胞，可释放抗体，溶解红细胞，导致溶血空斑的出现。通过计数溶血空斑的数量，可以判断活化 B 细胞的数量，通过测量每个空斑的大小，可以判断细胞产生抗体的多少。溶血空斑试验特异性高，筛选力强，可直接观察，故可用做判定免疫功能的指标。此处介绍的是其中的琼脂平板法。

【材料与试剂】

1. 试剂

20%SRBC 悬液，豚鼠血清，Hanks 液；底层琼脂（用 Hanks 液配成 1.4% 浓度）；上层琼脂（琼脂糖用 Hanks 液配成 0.4% 浓度，备用）。

2. 其他

平皿（直径 5.5cm × 1.5cm），镊子，100 目尼龙网纱布，温箱（37℃），水浴锅（45℃），离心机，显微镜及白细胞计数器，18 ~ 22g 小鼠等。

【操作方法】

1. 免疫脾细胞悬液的制备

（1）每只小鼠腹腔注入 SRBC 悬液 0.2ml，免疫后第 4 天进行后续试验。

（2）将小鼠拉颈处死，解剖取出脾脏，放入冷 Hanks 中漂洗后，置于 100 目尼龙网纱布上，加入少许 Hanks 液，用镊子挤压脾细胞，稍静置，吸上清液至离心管中，1500 转/分离心 5 分钟，弃上清后，以液 Hanks 液混匀，计数脾细胞数，调整细胞数至 10^7 个/ml。

2. 制备底层琼脂

将 1.4% 琼脂凝胶加热融化后，倾注于水平位置的平皿内，每皿 2 ~ 3ml。

3. 顶层琼脂的制备

将 0.4% 的琼脂糖融化后，置于 45℃ 恒温水浴箱中，依次加入 SRBC 悬液 0.1ml、豚鼠血清 0.5ml、10^7/ml 脾细胞悬液 0.1ml。迅速混匀后，倾注于已铺好底层琼脂的平皿内，使之均匀铺平凝固后，静置约 15 分钟，放 37℃ 温育 3 ~ 4 小时。

4. 计数 PFC

如需保存，可加入 0.25% 戊二醛 6ml 进行固定。

【结果与讨论】

1. 记录 PFC 数量。
2. 简述溶血空斑试验的过程和临床意义。

Ⅱ　接触性超敏反应

【实验目的】

1. 熟悉二硝基氟苯（DNFB）导致接触性超敏反应的原理。
2. 掌握 DNFB 引发接触性超敏反应的方法。

【基本原理】

接触性超敏反应是体内检测细胞介导的免疫功能的一种较为简便方法。在这个反应中，机体表皮暴露在曾经被致敏的外源性的半抗原下，将会导致迟发型超敏反应。这样的反应可定性，也可定量。

DNFB 可作为半抗原涂抹腹壁。待与皮肤内的组织蛋白结合，成为完全抗原，刺激 T 淋巴细胞增殖成为致敏淋巴细胞。经此抗原致敏的小鼠，在耳部涂抹 DNFB 后，可对 DNFB 产生特异的迟发型超敏反应，耳部出现肿胀。镜检可见大量的单核细胞，淋巴细胞浸润。

【实验材料】

小鼠两只，1% DNFB 溶液（需新鲜配制，用 1 份丙酮加 1 份麻油稀释），棉签，直径 8mm 打孔器，天平。

【实验方法】

1. 将两只小鼠腹部去毛，其中一只在腹部涂抹 1% DNFB100μl，另一只涂抹不含 DNFB 的丙酮麻油 100μl，做为对照。次日，可再强化免疫一次。

2. 5 天后，每只小鼠的右耳两面涂上 1% DNFB 溶液 10μl。

3. 24 小时后，处死小鼠。用打孔器在左右耳分别取下直径 8mm 的耳片，称重。同时取胸腺、脾脏称重。获得小鼠左右耳重量差，作为肿胀度。分别以每 10g 小鼠的脾重和胸腺重作为脾指数和胸腺指数，观察小鼠间差别。

【结果与讨论】

列表记录实验结果，比较致敏小鼠和未致敏小鼠的各项指标。

48 The Research of Immunomodulatory Drugs

The efficacy studies of immunomodulatory drugs have been generally used with a combination of *in vitro* and *in vivo* approach. *In vitro* studies can been used to determine which aspect of the immune response is affected by drugs. *In vivo* studies can explore the total effects of the drugs on the level of immune response, normal immune response and abnormal immune responses. Study of drugs on the immune system and immune function is not only reveals the mechanism of drugs, but also provide a basis for disease treatment.

The initial study of immunomodulatory drugs mainly contain four parts: the establishment of animal models, innate immune functions; acquired humoral immunity functions; acquired cellular immunity functions.

To evaluate the role of drugs, the selection of experimental animals, in addition to the normal animals, tend to select more than one type of animal models, such as immune suppression (for example the inhibitors, such as cyclophosphamide, hydrocortisone, inhibitors) or immuno - compromised (by radiation or other methods).

To evaluate the roles of drugs on innate immune, phagocytes, lysozyme *etc* are main research objectives. The related experimental methods can refer to the experimental 41, 42.

The roles of drugs on acquired humoral immune can be studied from many aspects. Single expansion, nephelometry can be used to determine immunoglobulin types, such as IgG, IgA, IgM, etc. To detect a variety of specific antibodies, serum hemolysin assay, ELISA and other methods can be used. To detect antibody producing cells, hemolytic plaque test, hemolysin spectrophotometry, *etc.* can be used. In this experiment, hemolytic plaque test are introduced.

To evaluate the roles of drugs on acquired cellular immune function, lymphocyte transformation test, contact allergic reactions are used. In this experiment, DNFB - induced allergic contact dermatitis are introduced.

Ⅰ Hemolytic Plaque Test

Purpose

1. To master the principle of hemolytic plaque test.
2. To master the methods of hemolytic plaque test.

Principle

Hemolytic plaque test is used for detecting single B lymphocytes (antibody – forming cells) *in vitro*. After immuned by sheep red blood cells (SRBC), B lymphocytes can release antibodies, dissolve red blood cell, result the emergence of plaque. Through counting the number of plaque, the number of activated B cells can been determine. Through measuring the size of each plaque, how many antibodies are produced by cells can been determine. With high specificity, better bolting capacity, observed directly, hemolytic plaque test can be done to determine the immune function. Described here is one of the agar plate method.

Materials and Apparatus

1. Reagent

20% SRBC, serum of guinea pig, Hanks solution, bottom agar (prepared with Hanks to concentration of 1.4%), top agar (agarose prepared with Hanks to concentration of 0.4%).

2. Others

Petri dish (diameter 5.5cm × 1.5cm), tweezer, 100 mesh nylon gauze, incubator (37 ℃), water bath pot (45℃), centrifuge, microscopes and interleukin counter, mouse weigh 18 ~ 22g.

Procedures

1. Preparation of immuned spleen cell suspension

(1) Immune each mouse with 0.2ml SRBC suspension through intraperitoneal injection. After four days, continue the test.

(2) The mice will be killed by pulling neck. Spleen will removed, put in cold Hanks after rinsing, and then placed in 100 mesh nylon gauze. After add a little Hanks solution, squeezed splenocyte with tweezers. Standing for a while, then remove the supernatant to centrifuge tubes, 1500r/min centrifugal 5min. Discard the supernatant, then mix with Hanks. Counting the number of spleen cells, adjust cell concentration to 10^7 cells / ml.

2. Preparation of bottom agar

1.4% agarose gel will be melted after heating, poured into the plate in the horizontal position for 2 ~ 3ml per dish.

3. Preparation of top agar

0.4% agarose gel will be melted after heating, put into 45℃ water bath. After by adding SRBC suspension 0.1ml, guinea pig serum 0.5ml, 10^7/ml spleen cell suspension 0.1ml, rapid mixing it and pour into the bottom agar. Standing about 15min, and incubated 37 ℃ for 3 ~ 4h.

4. Count PFC

For preservation, the plate can be added 0.25% glutaraldehyde to fix 6ml.

Results and Discussion

1. Record the number of PFC.

2. Briefly describe the process and clinical significance of Hemolytic plaque test.

Ⅱ Contact Hypersensitivity

Purpose

1. To be familiar with the principles of the Contact hypersensitivity induced by dinitrofluorobenzene (DNFB).

2. Master the method to induce contact hypersensitivity with DNFB.

Principles

Contact hypersensitivity is a simple *in vivo* assay of cell – mediated immune function. In this procedure, exposure of epidermal cells to exogenous haptens which they have previously been exposed results in a delayed – type hypersensitive reaction that can be measured and quantified.

DNFB can be used as hapten to daub abdominal wall. If DNFB combined with proteins in skin, it will become to a complete antigen and can stimulate T lymphocyte proliferation to sensitized lymphocytes. After sensitized to this antigen, daubing DNFB on the ear, mice can have a DNFB – specific delayed – type hypersensitivity and the ear will swelling. Examination with microscope will show a large number of mononuclear cells and lymphocyte infiltration.

Materials and Apparatus

Two mice, 1% DNFB solution (to be prepared freshly with 1 plus acetone sesame oil dilution), cotton, 8mm diameter hole puncher, the balance.

Procedures

1. Remove the abdomen fur of two mice. One of which smear 1% DNFB 100μl on the abdomen, the other applying DNFB free acetone oil 100μl, as a control. The next day, they can be immunized again.

2. 5 days later, each mouse's right ear on both sides painted with 1% DNFB solution of 10μl.

3. 24 hours later, the mice were killed. To remove the ear piece by about 8mm in diameter with the hole puncher. At the same time take the thymus, spleen weight. The weight difference between left and right ear pieces is the swelling degree. The spleen index and thymus index are spleen weight and thymus weight in 10g mice weight. Observe the difference between two mice.

Results and Discussion

Record the results, compare indicators between sensitized mice and sensitized mice.

第八部分　药学微生物学实验

实验四十九　土壤中抗生素产生菌的分离

【实验目的】

1. 了解从土壤中分离与纯化放线菌的基本原理及常用方法。
2. 掌握抗生素产生菌的抗菌谱测定方法。

【实验原理】

放线菌是重要的抗生素产生菌，许多临床应用的抗生素都是由土壤中分离的放线菌产生的。放线菌一般在中性偏碱性、有机质丰富、通气性好的土壤中含量较多。由于土壤中的微生物是各种不同种类微生物的混合体，必须把各种放线菌从这些混杂的微生物群体中分离出来，从而获得某一放线菌的纯菌株。根据放线菌对营养、酸碱度等条件要求，常选用合成培养基或有机氮培养基，也可采用选择性培养基或加入某种抑制剂，使细菌、霉菌出现的数量大大减少，从而分离土壤中的放线菌；再通过稀释法，使放线菌在固体培养基上形成单独菌落，并可得到纯菌株。

抗生素是放线菌的次级代谢产物，放线菌经液体培养后，其分泌的抗生素存在于离心所得的上清液中，可采用微生物方法进行检测，从而筛选到所需的抗生素产生菌。

【实验材料】

1. 土壤

校园土，空气干燥，磨碎。

2. 试验菌

金黄色葡萄球菌和大肠埃希菌的 8 小时肉汤培养物。

3. 培养基

淀粉培养基，高氏 1 号合成培养基、营养琼脂培养基。

4. 其他

镊子，药敏试纸片，灭菌的生理盐水，无菌平皿，无菌玻璃铲，无菌涂布棒，无菌移液管等。

【实验方法】

（一）土壤中放线菌的分离

1. 采集土样

土壤的种类和自然条件影响着放线菌的种类和分布数量。链霉菌主要存在于干燥、偏碱而营养丰富的土壤里；而小单孢菌多分布在潮湿土壤或湖底泥土中。因此在采土时，应注意地区、时间和植被情况。采土季节以春秋二季为宜，雨季不宜采土。一般认为，南方地区的土壤和北方地区土壤相比，南方地区的土壤链霉菌的种类较多，而某些特殊土壤或地区也可能存在一些特殊菌种。采集的土样最好随即分离，否则要放在阴凉通风处，防止变潮生霉。

在选定的采土地点，用小铲除去5cm厚表层土，用酒精棉球擦拭过的铁锹深挖15～20cm，采土后放入一个无菌纸袋或培养皿内，并按要求记录有关土样的内容，包括土壤编号、采集日期、采集地点及土壤基本特征。

2. 土壤悬液梯度稀释

（1）将土样放入用酒精擦拭过的乳钵中，除去石块、草根，研磨压碎后，称取5.0g放入盛有50ml灭菌的生理盐水的三角瓶中。振荡5分钟，即10^{-1}的土壤悬液，静置30秒。

（2）取1 ml土壤悬液，加入到9 ml灭菌生理盐水中10倍稀释。

（3）按1∶10梯度稀释至10^{-4}、10^{-5}和10^{-6}。

3. 制备培养基平板

（1）准备9个无菌玻璃平皿，分别标记10^{-4}、10^{-5}和10^{-6}，每个稀释度平行3皿。

（2）配制高氏1号琼脂培养基并于115℃灭菌20分钟，待其冷却至50～55℃后，将其倒入无菌平皿中，每个平皿加15～20ml，待冷凝成平板。

土壤放线菌的分离见图49－1。

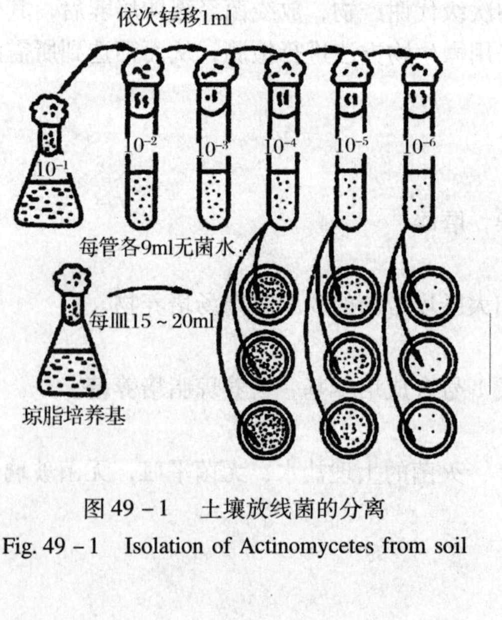

图49－1　土壤放线菌的分离

Fig. 49－1　Isolation of Actinomycetes from soil

4. 分离培养

从对应稀释度的土壤悬液中分别吸取0.5ml稀释液加到冷凝好的高氏1号平板中，用无菌涂布棒将加在平板培养基上的土壤稀释液在整个平板表面涂匀（每个稀释度换一支无菌涂布棒）。平板倒置于培养箱28℃恒温培养1周。

5. 将平板上的放线菌菌落分别单菌落挑取，接种于高氏1号琼脂培养基斜面，获得放线菌的纯培养菌株，28℃恒温培养1周，用于观察放线菌生长特征和产抗生素能力的测定。

（二）移块法粗筛抗生素（图49－2）

1. 在灭菌的空平皿中加入金黄色葡萄球菌肉汤培养液4～5滴，倒入融化并冷却到50℃的营养琼脂培养基20ml，制作混菌平板。

2. 在培养基表面贴上$1cm^2$的放线菌琼脂块5块（中间为已知有抑菌作用的链霉菌块作为阳性对照，周围4个为未知琼脂菌块，可以用从土壤中分离到的放线菌块）。

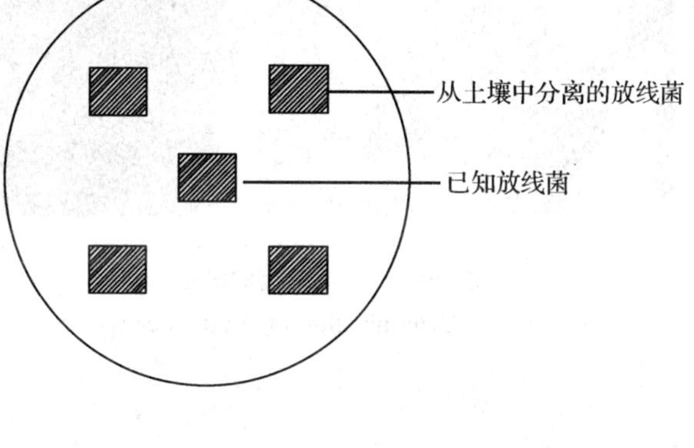

图49－2　移块法

Fig. 49－2 The moving－agar block method

3. 37℃，培养48小时，判定结果。如样品有抑菌作用，可在其周围出现不长菌的抑菌圈。

（三）抗菌谱测定（图49－3）

1. 发酵培养

（1）配制高氏1号合成培养基，分装25ml于250ml三角摇瓶中（每株放线菌对应1瓶），115℃灭菌20分钟。

（2）将分离到的每一株放线菌分别接种于摇瓶中，28℃、240转/分恒温培养1周。

（3）发酵液过滤，滤液用于抗菌活性测定。

2. 抗菌活性测定

（1）将10ml灭菌的营养琼脂培养基加入到已灭菌的平皿中、冷却，制备底层平板。

（2）取培养8小时的金黄色葡萄球菌和大肠埃希菌液体培养物2ml分别加到200ml灭菌的营养琼脂培养基中（注意培养基应冷却到50～55℃），震荡混匀，吸取6ml加入到底层平板上，分别制备含金黄色葡萄球菌和大肠埃希菌的双层平板。

（3）用镊子将一系列吸附有放线菌发酵液的无菌滤纸片置于金黄色葡萄球菌和大肠埃希菌的双层平板上，37℃培养 24 小时。

（4）观察抑菌圈，判断放线菌是否产生抗生素。

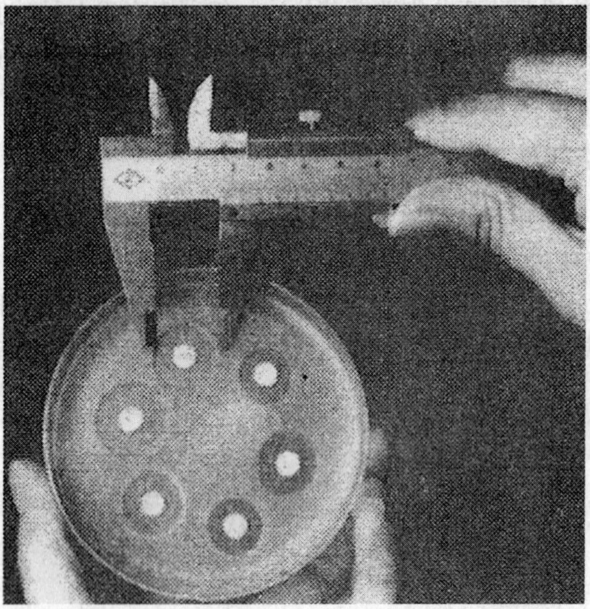

图 49－3　抗菌谱的测定

Fig. 49－3　Determination of Antibiotics spectru

【结果与讨论】

1. 描述从土壤中分离的放线菌菌落的形态特征。
2. 描述分离的放线菌所产生的抗生素的抗菌谱。
3. 在分离土壤中抗生素产生菌时，如何避免细菌和真菌的生长？

PART EIGHT General Microbial Experiments in Pharmaceutics

49 Isolation of Antibiotic – Producing Actinomycetes from Soil

Purpose

1. To learn the basic principles and the common methods to isolate and purify actinomycetes from soil.
2. To be familiar with the determination of anti – bacterial spectrum of antibiotics.

Principle

Actinomycetes are important antibiotic – producing bacteria. A lot of useful antibiotics in clinic are obtained by actinomycetes which are isolated from soil. They are usually found in the soil with neutral or alkaline pH, adequate organics and good ventilation. Since the microorganisms in the soil are always mixture of different species, it is necessary for us to isolate actinomycetes of different kinds from the colony of microorganisms mixture to get certain pure culture of actinomycetes. Synthetic medium or organic nitrogen culture is commonly used to meet the requirement of actinomycetes, alternatively, we can use selective culture media or add certain inhibitor to reduce the number of bacteria and molds substantially, and then we can isolate actinomycetes from soil. After that, we can get individual colony of actinomycetes on solid culture media by dilution, and then obtain pure strain.

Antibiotics are the secondary metabolites of actinomycetes. After fermented in liquid media and centrifuged, antibiotics in the supernatant secreted by antibiotics – producing actinomycetes can be determined with microbial method and screen the antibiotic – producing bacteria.

Materials and Apparatus

1. Soil

Soil obtained from campus by air dry and powdering.

2. Cultures

8h broth culture of *S. aureus* and *E. coli.*

3. Culture media

Starch media, starch agar media, Gauss' No. 1 deffined media, nutrient agar media.

4. Others

Tweezers, paper slips, sterile physiological saline, incubator, etc.

Procedures

A. Isolation of Actinomycetes from Soil

1. Collect sample soil

The species and distribution of actinomycetes are affected by species and natural conditions of soil. Streptomycin mainly exists in dry, alkaline and nutrient – riched soil, while micromonospora mainly distributes in damp soil or lake – bottom clay. Therefore, region, time and condition of vegetation should be considered when sampling soil. It is appropriate in spring and autumn, but not in rainy days. The species of actinomycetes in the South are more than those of the North. Certain special soil or region may have some special species of actinomycetes. It is better to isolate the strains from the sample soil right after sampling, or preserve the soil in cool places to prevent mildewing.

In selected region, remove the surface soil of 5cm depth with a shovel and then dig to 15 ~ 20 cm deep with alcohol cleaned spade. Put the collected soil in a sterile paper bag or plate, and record the numbers. with sample name, date, place of sampling and the basic characteristic of the soil.

2. Prepare a serial dilution of the soil samples as follows

(1) Put the sample soil to mortar which is cleaned with alcohol, remove stones, roots of grass, levigate the soil. Weigh 5. 0g soil and put it into the flask containing 50 ml sterile physiological saline. Mix thoroughly for 5 min to make a uniform soil – water suspension.

(2) Transfer 1ml soil suspension to 9ml sterile physiological saline for 10 times dilution.

(3) Repeat the 1 : 10 dilution to obtain 10^{-4}, 10^{-5} and 10^{-6} diluted solution.

3. Prepare culture media plates

(1) Prepare 9 sterile glass plates, label three sets of sterile plate with the dilutions (10^{-4}, 10^{-5} and 10^{-6}).

(2) Make Gauss'No. 1 agar media and sterilize at 115℃ for 20 min, cool to 50 ~ 55℃, add 15 ~ 20ml of it to each plate and curdle.

4. Isolate and incubate

Transfer 0. 5 ml diluted solution of 10^{-4}, 10^{-5} and 10^{-6} to three sterile plates respectively and spead it. Incubate all plates in an inverted position at 28℃ for one week.

5. Aseptically isolate actinomycetes colonies to starch agar plate respectively and incubate at 28℃ for one week for the observation growth feature of actinomycetes and the determination of antibiotics.

B. The moving – agar block method

This method is applicable to test disinfectant or inhibitory ability of ointment or agar strain, and it can be used to preliminarily screen actinomycetes isolated from soil and its antibiotic – producing antibiotic ability.

1. Add 4 ~ 5 drops of broth culture of *S. aureus* to an empty plate, and then pour 15 ~ 20 ml of agar medium which has been melt and then cooled to 50℃. Mix well.

2. Attach the surface of the culture medium with five 1cm^2 size of agar blocks of actinomycetes isolated from soil. (figure 49 – 1)

3. Incubate at 37℃ for 48h and then observe the inhibitory circle.

C. Determination of Antibiotics Spectrum

1. Fermentation and incubation

(1) For each actinomycetes, prepare starch media and pour 25ml to a 250 ml shaking flask, sterilize at 115℃ for 20 min.

(2) Inoculating agar slant cultures of actinomycetes to each shaking flask, in 28℃ 240r/min incubator for a week.

(3) Filtrate the fermentation broth for the determination of antimicrobial activity.

2. Determination of antimicrobial activity

(1) Pour 10ml sterile nutrient agar media to a sterile plate and cool to make bottom plate.

(2) Transfer 2ml of 8h cultures of *S. aureus* and *E. coli* to 200ml sterilized nutrient agar media (50 ~ 55℃) respectively. For each bacterium, pour 6ml to the upper nutrient agar plate to prepare two layers agar plate.

(3) Add a series of paper slip absorbed the fermentation broth on the two layers agar plate for *S. aureus* and *E. coli*, incubate at 37℃ for 24h.

(4) Observe the inhibitory circle and determine whether actinomycetes can produce antibiotics.

Results and Discussion

1. Describe morphological feature of Actinomycetes colonies isolated from the soil.

2. Describe antibiotic spectrum produced by Actinomycetes isolated from the soil.

3. How to avoid the growth of bacteria and fungi when isolating antibiotic – producing actinomycetes from the soil?

实验五十　药物的体外抗菌试验

【实验目的】

1. 熟悉常用的药物体外抗菌活性的测定方法。

2. 了解平板倍比稀释法和试管液体稀释法测定抗生素的最低抑菌浓度（MIC）的方法。

【实验原理】

药物体外抗菌活性的测定方法很多，一般有两大类：琼脂扩散渗透法和系列浓度稀释法。广泛应用于新药研究和指导临床用药，如抗菌药物筛选、药物的抗菌谱测定、药敏试验、药物血浓度测定等。

琼脂扩散渗透法是利用药物能够渗透到琼脂培养基的性质，将试验菌混入琼脂培养基后倾注倒平板，或将试验菌涂布于琼脂平板的表面，然后用不同的方法将药物置于已含试验菌的琼脂平板上。根据加药的操作方法不同分为滤纸片法、打洞法、挖沟法、管碟法和移块法等，经适宜温度培养后观察药物的抑菌能力。本实验介绍滤纸片法和挖沟法。

系列浓度稀释法常用于测定药物的最低抑菌浓度（minimal inhibitory concentration，MIC）或最低杀菌浓度（minimal bactericidal concentration，MBC）。药物的最低抑菌浓度是指药物能够抑制微生物生长的最低浓度；可杀灭细菌的最低药物浓度为最低杀菌浓度。MIC（MBC）可以评价药物抑（杀）菌作用的程度，常以（mg/ml 或 μg/ml 表示。其值愈小，则抑（杀）菌作用愈强。常见的测定方法有：试管稀释法、平板稀释法、斜面混入法和微孔板法等。本试验介绍平板稀释法、试管稀释法和微孔板法。平板倍比稀释法是将药物倍比稀释后，定量与琼脂培养基混合制备含药平板，再将细菌接种在含有不同浓度抗生素的平板上恒温培养，抑制细菌生长的最低药物浓度为最低抑菌浓度。试管稀释法是在一系列试管中，用液体培养基按照几何级数或数学级数连续稀释药物，然后在每一试管中加入一定量的试验菌培养，经培养后，肉眼观察能抑制试验菌生长的最低浓度即为该药物的 MIC。

【材料和仪器】

1. 菌株

金黄色葡萄球菌、表皮葡萄球菌、大肠埃希菌和肺炎克雷伯菌临床菌株的 8 小时肉汤培养物，金黄色葡萄球菌（ATCC25925）、大肠埃希菌（ATCC25922）标准菌株的 8 小时肉汤培养物。

2. 待测药物

青霉素 G（β－内酰胺类）、链霉素（氨基糖苷类）、阿奇霉素（大环内酯类）和左旋氧氟沙星（喹诺酮类）、0.1% 新洁尔灭、0.1% 龙胆紫、2.5% 碘液、0.85% 氯化钠溶液。

3. 培养基

灭菌并恒温于55℃水浴的M－H（Mueller－Hinton agar）培养基、营养琼脂培养基。

4. 试剂和仪器

无菌生理盐水，无菌平皿，无菌吸管，无菌试管，无菌铲，无菌滤纸片，接种环，镊子等。

【实验方法】

（一）滤纸片法

滤纸片法是琼脂扩散法中最常用的方法，适用于新药的初筛试验（初步判断药物是否有抗菌作用）及临床的药敏试验（细菌对药物的敏感性试验，以便临床选择治疗用药参考）。可进行多种药物或一种药物的不同浓度对同一种试验菌的抗菌试验。

以一定直径（6～8mm）的无菌滤纸片，蘸取一定浓度的被检药液，将其紧贴在含菌平板上，如果纸片上含有药液，便会沿琼脂向四周扩散，且对该试验菌有抑制作用，经一定时间培养后，就可在滤纸片周围形成不长菌的透明圈。见图50－1。

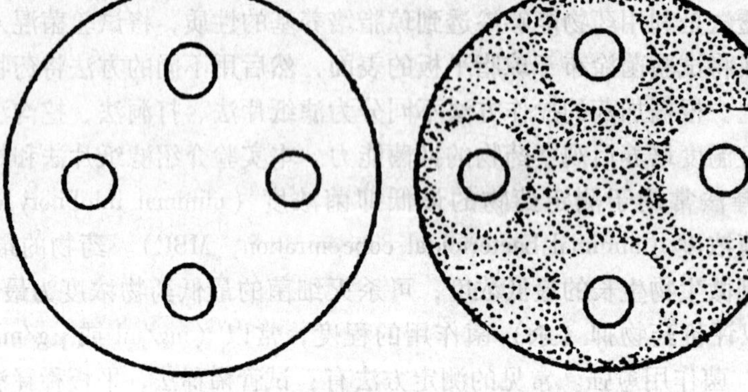

培养前　　培养后

图50－1　滤纸片法

Fig. 50－1　The paper－disk diffusion test

1. 用滴管分别取金黄色葡萄球菌和大肠埃希菌肉汤培养物4～5滴，加到两个灭菌的空平皿中，每皿加入20ml已溶化并冷却至50℃左右的培养基，制成含菌平板，冷凝备用。

2. 用无菌镊子夹取滤纸片，分别浸入0.1%新洁尔灭、0.1%龙胆紫、2.5%碘液、0.85%生理盐水中。在盛药平皿内壁上除去多余药液后，分别贴在含菌平板表面，并作好标记，37℃，培养20小时。

3. 观察滤纸片周围的抑菌圈。滤纸片边缘到抑菌圈边缘的距离在1mm以上者为阳性（＋），即微生物对药物敏感；反之为阴性（－）。

（二）挖沟法

本法适用于半流动药物或中药浸煮剂的抗菌试验。可在同一平板上试验一种药物对几种试验菌的抗菌作用。

1. 在琼脂平板中央，用无菌铲挖一条长沟，将沟内琼脂全部挖出。

2. 将待测药物加入此沟内，以装满不流出为限。

3. 在沟两侧垂直划线接种各种试验菌。见图 50－2。

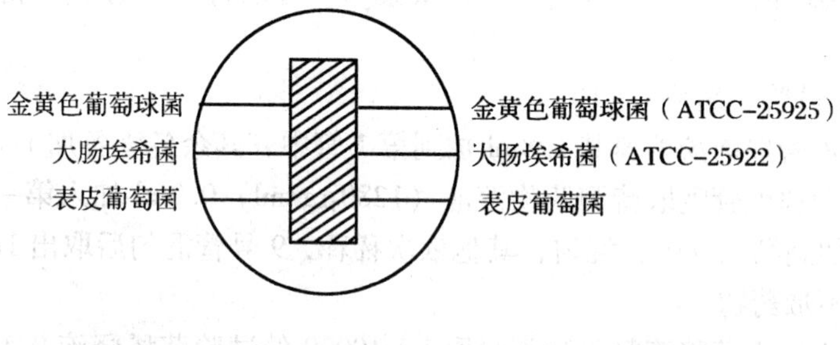

图 50－2　挖沟法

Fig. 50－2　The trenching test

4. 若为细菌则 37℃培养 24～48 小时；若为放线菌或真菌则 28℃培养 48～72 小时。

5. 观察沟两边所生长的试验菌离沟的抑菌距离，从而判断待测药物对这些菌的抗菌能力。

（三）平板倍比稀释法

1. 药物配制

（1）称取适量抗生素，用灭菌生理盐水配制成浓度为 1280μg/ml 的溶液。

（2）用灭菌生理盐水倍比稀释成浓度为 0、20、40、80、160、320、640、1280μg/ml 的溶液。

2. 含药物平板的制备

（1）分别吸取上述各种浓度的药物溶液 1.5ml 于无菌平皿中。

（2）加入 13.5ml 恒温于 55℃的 M－H 培养基，迅速混匀、冷却制成一系列药物浓度为 0、2、4、8、16、32、64、128μg/ml 的药物平板（药物浓度为 0 的平板作阴性对照）。一般配制好的含药琼脂平板应装入密封塑料袋中，置 2～8℃冰箱可贮存 5 天。

3. 试验菌与点样

（1）将金黄色葡萄球菌、表皮葡萄球菌、大肠杆菌和肺炎克雷伯氏菌、金黄色葡萄球菌（ATCC25925）、大肠杆菌（ATCC25922）的 8 小时肉汤培养物用生理盐水 1：10000 稀释，细菌浓度为 10^7～10^8CFU/ml。

（2）用笔将药物平板分为 6 个区，标记菌名，分别用接种环将试验菌对应点种于系列药物平板上，接种量约为 10^4～10^5CFU/点，形成直径为 5～8mm 的菌斑。

4. 培养与结果观察

（1）37℃恒温培养 16～18 小时。

（2）观察药物对测试菌生长的抑制作用，以抑制细菌生长的最低药物浓度记录为MIC 值。

（四）试管稀释法

该方法是在一系列试管中，用液体培养基按照几何级数或数学级数连续稀释药物，然后在每一试管中加入一定量的试验菌，经培养后，肉眼观察能抑制试验菌生长的最低浓度即为该药物的 MIC。本法操作中所用药物及试验菌与平板倍比稀释法相同，选做其中一种菌——金黄色葡萄球菌，其中一种药——青霉素 G。可以将两种方法所得结果数据进行比较。

1. 取 10 支小试管，编号 1 ~ 10。

2. 用 5ml 移液管取肉汤培养基 1. 8ml 加到第 1 管中，其余各管各加 1ml。

3. 用 1ml 无菌移液管吸取待测药物溶液（1280μg/ml）0. 2ml 加入第一管内混匀，从第一管取出 1ml 加到第二管内，混匀，其他依次稀释，9 号管混匀后取出 1ml 扔掉，10 号管为空白对照，不加药物。

4. 用另一支 1ml 无菌移液管，分别吸取 1∶10000 的试验菌稀释液 0. 1ml 加到含有不同浓度药液和对照小试管中，加入顺序为从对照管（10 号）开始依次从药液浓度低向药液浓度高的试管进行。稀释过程见表 50 – 1。

5. 37℃，培养 20 小时，对照管中菌应正常生长，液体变为混浊。观察其他管中药物对测试菌生长的抑制作用，以抑制细菌生长的最低药物浓度记录为 MIC 值。

表 50 – 1　试管稀释法系列稀释过程

管号	1	2	3	4	5	6	7	8	9	10
肉汤培养基（ml）	1. 8	1. 0	1. 0	1. 0	1. 0	1. 0	1. 0	1. 0	1. 0	1. 0
药液（ml）	0. 2 →	1. 0→	1. 0→	1. 0→	1. 0→	1. 0→	1. 0→	1. 0→	1. 0→ 弃掉	
试验菌（ml）	0. 1	0. 1	0. 1	0. 1	0. 1	0. 1	0. 1	0. 1	0. 1	0. 1
药物终浓度（μg/ml）	128	64	32	16	8	4	2	1	0. 5	0
每管总体积（ml）	1. 1	1. 1	1. 1	1. 1	1. 1	1. 1	1. 1	1. 1	1. 1	1. 1

（五）微孔板法

此法基本同试管稀释法，不同之处是药液稀释后，用微量吸管加入不同孔中，每孔加100μl 药物稀释液，然后再加 10μl 菌液，混匀，在适当的温度下培养。微孔板是一种聚氯乙烯塑料板，通常用 96 孔等，可用于同时测定多种药物的最低抑菌浓度。

【结果与讨论】

1. 各种化学药品对金黄色葡萄球菌的抑菌作用。

将实验结果填入表 50 – 2。

表 50-2　滤纸片法抑菌结果

化学药品	抑菌圈直径（mm）
0.1%新洁尔灭	
0.1%1龙胆紫	
2.5%碘液	
0.85%氯化钠溶液	

2. 观察挖沟法实验中沟两边所生长的试验菌离沟的抑菌距离，判断药物对这些试验菌的抑菌能力大小。

3. 观察平板倍比稀释法的结果，填入表 50-3。

表 50-3　平板稀释法 MIC 结果

菌名	药物	药物浓度（μg/ml）									IC
		128	64	32	16	8	4	2	1	0	（μg/ml）
金黄色葡萄球菌	青霉素 G										
	链霉素										
	阿奇霉素										
	左旋氧氟沙星										
金葡 ATCC25925	青霉素 G										
	链霉素										
	阿奇霉素										
	左旋氧氟沙星										
表皮葡萄球菌	青霉素 G										
	链霉素										
	阿奇霉素										
	左旋氧氟沙星										
大肠埃希菌	青霉素 G										
	链霉素										
	阿奇霉素										
	左旋氧氟沙星										
大肠 ATCC25922	青霉素 G										
	链霉素										
	阿奇霉素										
	左旋氧氟沙星										
肺炎克雷伯菌	青霉素 G										
	链霉素										
	阿奇霉素										
	左旋氧氟沙星 *										

生长，以“+”表示；不生长，以“-”表示。

4. 观察试管稀释法实验中各试管菌液的生长状况，记录第几号管液体刚好澄清，即能完全抑制试验菌生长的最高稀释倍数管，其抗生素浓度即为最低抑菌浓度 MIC。

5. 在本实验中，化学药物对微生物所形成的抑菌圈是否说明微生物已经被杀死？如何通过实验加以确定，请自行设计实验。如何判断试验中导致细菌不生长的药物浓度是抑菌还是杀菌？MIC 大，还是 MBC 大？

50 Determination of Antimicrobial Activity *in vitro*

Purpose

1. To be familiar with the common methods of determination of antimicrobial activity in vitro.

2. To understand the principles for the determination methods of the minimal inhibitory concentration (MIC) by double dilution on agar plate and series dilution in test tube.

Principle

There are many methods to determine antimicrobial activity *in vitro*, they are mainly classified as: the disk diffusion test and the serial dilution test. They are widely used in the research of new drug and clinical treatment, such as screening of antibiotics, determination of antimicrobial drug spectrum, susceptibility of antibiotics and measurement of drug concentrations in the blood.

The principal behind the disk diffusion test is as follows: when the antibiotic is placed on agar by some ways previously inoculated with the test bacteria, the antibiotic diffuses radially outward through the agar, producing an antibiotic concentration gradient. A clear zone or ring is present around the antibiotic after incubation if the agent inhibits bacterial growth. The wider the zone surrounding the antibiotic, the more susceptible the bacterium is. It is divided into the paper - disk diffusion test, the trenching test, the cup - plate diffusion test and the moving - agar block test. We introduce the paper - disk diffusion test and the trenching test.

The serial dilution tests usually used to determine drug minimal inhibitory concentration (MIC) and minimal bactericidal concentration (MBC). MIC is the minimum concentration of a drug that can inhibit the growth of microorganisms; MBC is the minimum concentration of a drug that can kill the microorganisms. We can assess the extent of drug inhibiting bacteria by MIC (MBC) and is usually represented by (mg/ml or μg/ml. The smaller MIC and MBC are, the higher antimicrobial activities are. The commonly used methods are disk dilution test, broth dilution test, the infiltrating - agar slant test, the microwell plate test and so on. We introduce the disk dilution test and the broth dilution test. The disk dilution test is to dilute the drug 2 fold to proportional extent in an agar media plate. The test bacteria are inoculated on the agar plate containing series antibiotic concentrations. The lowest concentration of antibiotic resulting in inhibition and no growth after 16 to 20h of incubation is the minimal inhibitory concentration (MIC) and minimal bactericidal concentration (MBC) respectively. The broth dilution test is to continuously dilute drugs with broth medium based on geometric or mathematic series in test tubes, and then add a certain amount of test bacteria culture to each tube. After incubation, the minimum concentration of drugs that can inhibit the growth of test bacteria with naked eyes is defined as MIC.

Materials and Apparatus

1. Cultures

8h broth cultures of clinical strains of *Staphylococcus aureus*, *Staphylococcus epidermidis*, *Escherichia coli*, *Klebsiella pneumoniae*, and standard strains of *Staphylococcus aureus* (ATCC25925) and *Escherichia coli* (ATCC25922) .

2. Drugs to be determined

Penicillin G (β – lactam), Streptomycin (aminoglycosides), Azithromycin (macrolides) and Levofloxacin (quinolones), 0.1% Benzalkonium Bromide, 0.1% Gentian Violet, 2.5% iodine solution, 0.85% physiological saline.

3. Culture media

Sterilized M – H (Mueller – Hinton agar) culture medium at 55℃ in constant temperature bath, nutrient agar media.

4. Reagents and apparatus

Sterile physiological saline, sterile plates, sterile pipettes, sterile test tubes, sterile paper slips, sterile glass spade and inoculating loop.

Procedures

A. The paper – disk diffusion test

This is the most commonly used method in agar diffusion and is applicable to preliminarily screen for new drugs (to evaluate whether the drug has the antimicrobial effect) and to test clinical drug susceptibility. It also can be used to carry out antimicrobial experiment a test bacterium which is subjected to many drugs or one drug of different concentrations.

1. Transfer 4 ~ 5 drops of *S. aureus* and *E. coli* broth culture respectively to two sterile empty plates, pour 15 ~ 20 ml nutrient agar medium (50℃) for each plate and mix by gentle rotation.

2. Add a series paper absorbed 0.1% Benzalkonium Bromide, 0.1 gentian vio-

let, 2.5% iodine solution, 0.85% physiological saline respectively on the plates, and then label them. Incubate at 37℃ for 20h.

3. Observe the inhibitory bacteria circle around the paper. If the distance between rim of the filter paper and the inhibitory bacteria circle is over 1 mm, it is a positive result, which means the microorganisms are sensitive to the drug, otherwise it will be negative.

B. The trenching test

This method is applicable to the antimicrobial test of semi – fluid or the soak of Chinese medicinal materials. It can examine the effect of one drug on several different test bacteria on the same agar plate.

1. Dig out all agar of a ditch in the center of the agar plate with a sterile spade.

2. Add the testing drug to the ditch to its full capacity. Note: do not spill over the ditch.

3. Inoculate a variety of test bacteria to both side of the ditch. (As follows Fig. 50 –2).

4. For bacteria, incubate for 24 ~ 48h at 37℃, for Actinomycetes or Fungi incubate for 48 ~ 72h at 28℃.

5. Observe the inhibitory bacterial distance between ditch and test bacteria on both sides of the ditch, and then assess antimicrobial ability of test drugs.

C. The disk dilution test

1. Preparation of drugs solution.

(1) Weigh each kind of antibiotics and dissolved in sterilized physiological saline to obtain the antibiotics concentrations of 1280μg/ml respectively.

(2) Dilute the antibiotics solutions with sterilized physiological saline to concentrations of 0, 10, 20, 40, 80, 160, 320, 640 and 1280μg/ml.

2. Preparation of drug plates

(1) Add 1.5 ml of different antibiotics diluted solutions into a series of sterile plates.

(2) Pour 13.5 ml Mueller – Hinton agar medium incubated at 55℃ in water bath into each plate and mix gently and rapidly to form a series of drug agar plate with antibiotics concentrations of 0, 1, 2, 4, 8, 16, 32, 64 and 128 μg/ml (use drug whose concentration is 0 as negative control).

3. Test microorganisms and inoculating

(1) Dilute the 8h broth cultures of each tested bacteria with the dilution rate of 1 : 10000 to obtain bacteria concentration of approximately 10^7 to 10^8 CFU/ml.

(2) Label the drug plates into six areas and inoculate 1 ~ 2μl test bacterium on each area. The amount of bacteria for each spot is approximately 10^4 to 10^5 CFU.

4. Incubation and observation

(1) Incubate the inoculated plate at 37℃ for 16 to 18h.

(2) Observe the antimicrobial activity of different antibiotics and record MIC.

D. The broth dilution test

In a series of test tubes, the testing drugs are doubly diluted with broth culture medium. Add a certain amount of test bacteria to each test tube and then incubate. After that, the minimum concentration of the drug that can inhibit the growth of bacteria observed by bare eyes is MIC. We select one species—*S. aureus* and one drug—penicillin G and compare the results obtained by the disk dilution test and the broth dilution test.

1. Take out 10 test tubes, label with 1 ~ 10.

2. Add 1.8 ml broth culture medium to No. 1 tube and add 1 ml broth culture medium to other tubes.

3. Add 0.2 ml test drug solution (1280μg/ml) to No. 1 tube, mix evenly, draw 1 ml from No. 1 tube and add it to the second tube for double times dilution, mix evenly, repeat the double dilution from No. 2 to No. 9 tube and throw the last 1 ml away. The No. 10 is the control, no need to add the drug.

4. Add 0.1 ml diluted solution of test bacteria whose concentration is 1 : 1000 to the ten tubes, start from lower drug concentration to higher concentration.

5. Incubate for 20h at 37℃. The bacteria in the control tube should grow well and the liquid becomes opaque. Observe the antimicrobial activity of test drug in other tubes and record MIC. (Table 50 – 1)

Table 50 – 1 The broth dilution test

No. of tube	1	2	3	4	5	6	7	8	9	10
Broth culture (ml)	1.8	1.0	1.0	1.0	1.0	1.0	1.0	1.0	1.0	1.0
Drug sample (ml) diluting	0.2 →	1.0→	1.0→	1.0→	1.0→	1.0→	1.0→	1.0→	1.0→ lost	
Staphylococcus aureus (ml)	0.1	0.1	0.1	0.1	0.1	0.1	0.1	0.1	0.1	0.1
Concentration of drug (μg/ml)	128	64	32	16	8	4	2	1	0.5	0
Total volum (ml)	1.1	1.1	1.1	1.1	1.1	1.1	1.1	1.1	1.1	1.1

E. The microwell plate test

This method is similar to the broth dilution test in tube. The difference is to add solution with the bacterial culture to the microwell instead of the tube. Add 100μl drug diluted solution and 10μl bacterial solution to each hole, mix evenly and then incubate at appropriate temperature. The microwell plate is a plastic plate made of polyethylene which has 96 holes. It can be used to determine the minimal inhibitory concentration of many drugs.

Results and Discussion

1. The effect of chemical drugs on *S. aureus*, fill in the following table 50 – 2.

Table 50 – 2 The result of diameter of inhibition zone

Chemical drugs	Diameter of inhibition zone (mm)
0.1% Benzalkonium Bromide	
0.1% Gentian Violet	
2.5% iodine solution	
0.85% physiological saline	

2. Observe the distance between ditch and test bacteria grown on both sides of the ditch, assess the inhibitory ability of the drugs.

3. Observe the results of the disk dilution test, record in table 50 – 3.

Table 50 – 3 MIC of drugs.

Strain	Drugs	Concentration of drug (μg/ml)									MIC
		128	64	32	16	8	4	2	1	0	(μg/ml)
S. aureus	Penicillin G										
	Streptomycin										
	Azithromycin										
	Levofloxacin										
S. aureus ATCC25925	Penicillin G										
	Streptomycin										
	Azithromycin										
	Levofloxacin										
S. epidermidis	Penicillin G										
	Streptomycin										
	Azithromycin										
	Levofloxacin										
E. coli	Penicillin G										
	Streptomycin										
	Azithromycin										
	Levofloxacin										
E. coli (ATCC25922)	Penicillin G										
	Streptomycin										
	Azithromycin										
	Levofloxacin										
K. pneumoniae	Penicillin G										
	Streptomycin										
	Azithromycin										
	Levofloxacin										

Growth, represented by "+", no growth represented by "–"

4. Observe the growth of bacteria solution in each tube, record MIC.

5. In this experiment, ther is no growth of bacteria in the inhibition areas, is it concluded that those bacteria are killed? How to prove it? And design another experiment to examine whether the drug exhibite inhibitory or bacteriacidal effect? Which value is larger between MIC or MBC?

实验五十一 抗生素效价测定

【实验目的】

1. 熟悉微生物法测定抗生素效价的基本原理。
2. 了解管碟法（二剂量法）测定抗生素效价的基本原理。
3. 掌握用管碟法测定抗生素的效价。

【实验原理】

抗生素的效价常采用微生物学方法测定，它是利用抗生素对特定的微生物具有抗菌活性的原理来测定抗生素效价的方法，以抗生素的抑菌力和杀菌力作为衡量效价的标准，与临床应用具有平行相关性。微生物学法测定抗生素效价，一般可分为稀释法、比浊法和琼脂扩散法，以琼脂扩散法中的管碟法最为常用。

管碟法是根据抗生素在琼脂平板培养基中的扩散渗透作用，比较标准品和检品两者对试验菌的抑菌圈大小来测定检品的效价。基本原理是在含有高度敏感性试验菌的琼脂平板上放置小钢管［内径（6.0±0.1）mm，外径（8.0±0.1）mm，高（10±0.1）mm］，管内放入标准品和检品的溶液，经16~18小时恒温培养，抗生素扩散的有效范围内则产生透明的无菌生长的抑菌圈。抑菌圈的直径大小与抗生素的浓度相关，也与抗生素的扩散系数、扩散时间、培养基的厚度及抗生素的最低抑菌浓度等因素有关，比较抗生素标准品与检品的抑菌圈大小，可计算出抗生素的效价。管碟法的特点是灵敏度高，能直接显示抗生素的抗菌活性，因此作为国际通用的方法被列入各国药典中。

管碟法测定抗生素的效价又分为一剂量法、二剂量法和三剂量法。其中二剂量法最为常用，又称四点法。将抗生素标准品和检品各稀释成一定浓度比例（2：1或4：1）的两种溶液，在同一平板上比较其抗菌活性，再根据抗生素浓度对数和抑菌圈直径成直线关系的原理来计算检品效价。取含菌的双层平板培养基，每个平板表面放置4个小钢管，管内分别放入检品高、低剂量和标准品高、低剂量溶液。

先测量出四点的抑菌圈直径，按下列公式计算出检品的效价。

（1）求出W和V：

$$W = (SH + UH) - (SL + UL)$$
$$V = (UH + UL) - (SH + SL)$$

式中，UH：检品高剂量之抑菌圈直径；

UL：检品低剂量之抑菌圈直径；

SH：标准品高剂量之抑菌圈直径；

SL：标准品低剂量之抑菌圈直径。

（2）求出θ：

$$\theta = D \cdot \text{antilg}(IV/W)$$

式中，θ：检品和标准品的效价比；

D：标准品高剂量与检品高剂量之比，一般为 1；

I：高低剂量之比的对数，即 lg2 或 lg4。

（3）求出 Pr：

$$Pr = Ar \times \theta$$

式中，Pr：检品实际单位数；

Ar：检品标示量或估计单位。

二剂量法也可利用放线图，查出抗生素的效价，这样可以节省计算时间并便于核对。放线图系根据效价计算公式（θ = D · antilg（IV/W））推导而制得，只要求得 W、V 值后，查放线图即可得 θ 值（图 51－1）。

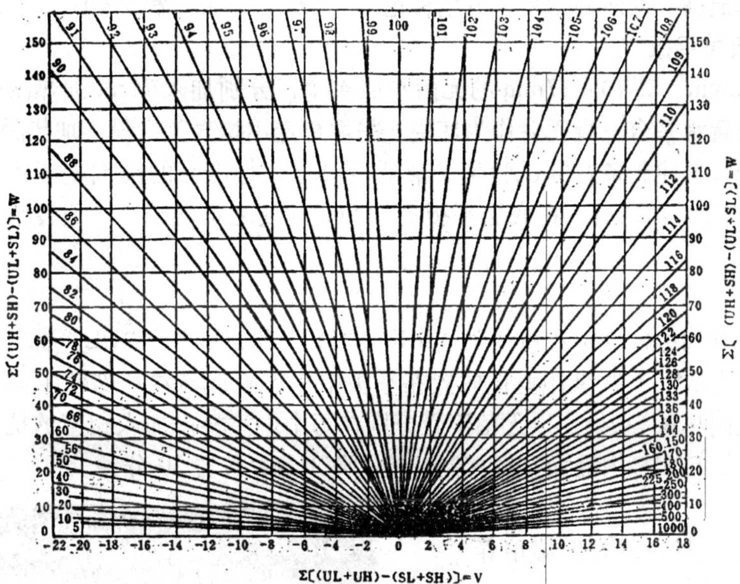

图 51－1　二剂量法效价计算放线图（H：L＝2：1）

Fig. 51－1　The figure of w，v and θ

药典中规定在测定抗生素效价时，对所用的标准品、试验菌、培养基、培养条件及药物的浓度范围都有相应的要求。本试验参照 2005 年版的药典选择测定四环素检品的效价。

【材料和仪器】

1. 菌株

藤黄微球菌［CMCC（B）28001］。

2. 培养基

效价检定用培养基配方为（g/L）胨：10；牛肉浸出粉：6；葡萄糖；1，琼脂：15～20，115℃灭菌 30 分钟。

3. 抗生素

四环素检品及标准品（高剂量、低剂量，高剂量与低剂量之比为 2：1，浓度范围：

10.0 ~ 40.0 U/mg)，无菌磷酸盐缓冲液 pH：6.0。

4. 试剂与器材

灭菌生理盐水，无菌平皿，牛津杯（小钢管），无菌陶土盖，无菌吸管，镊子，滴管，卡尺等。

【实验方法】

1. 制备试验菌悬液

取藤黄微球菌［CMCC（B）28001］接种于营养琼脂斜面，26 ~ 27℃培养 24 小时，用 0.9% 灭菌氯化钠溶液将菌苔洗下备用。

2. 供试品溶液的制备

按药典所要求药物浓度范围精密称取适量四环素标准品和检品，用无菌磷酸盐缓冲液溶解后，稀释成高、低两种剂量，分别标记为“SH”、“SL”和“UH”、“UL”备用。

3. 制备双层平板

取直径 90mm、高 16 ~ 17mm 的无菌平皿 4 个，分别加入 20ml 加热融化的无菌检定用培养基，放置水平台上凝固后作为底层。另取检定用培养基适量，加热融化后，冷却至 50℃左右，加入上述制备好的试验菌悬液适量（以使标准品的高剂量所致抑菌圈直径在 18 ~ 22mm 为宜），摇匀后，在底层平板上加 5ml 并使之均匀摊平，放置水平台上凝固后作为含菌层。

4. 检定

在平板底部四边，分别对角注明“SH”、“SL”和“UH”、“UL”标记。镊子火焰灭菌 3 次，然后夹取 4 个牛津杯垂直放置在平板中标志附近（注：勿使牛津杯陷入培养基内，各小杯之间尽量等距）。以无菌滴管分别在每个牛津杯内加入相应的抗生素溶液至满但不溢出管外，且四杯内液面高度相同，换上陶土盖，静置 30 分钟。

5. 培养

将平皿置于 35 ~ 37℃恒温培养箱培养 16 ~ 18 小时。

6. 观察结果

用玻璃盖换下陶土盖，倒出牛津杯，用卡尺量取各抑菌圈直径（SH、SL、UH、UL），以毫米为单位，4 个平板中误差不超过 0.1mm，代入公式计算检品效价。

见图 51 - 2。

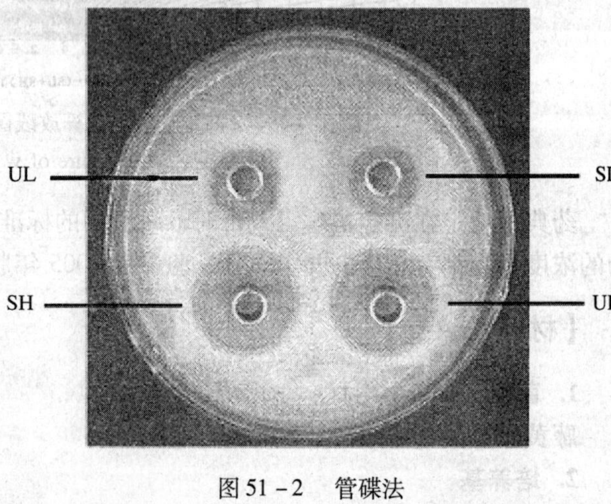

图 51 - 2 管碟法

Fig. 51 - 2 The cylinder plate method

【结果与讨论】

1. 记录抑菌圈直径并计算抗生素效价（已知四环素标准品 1000U/mg）。

表 51－1　抑菌圈测量结果

碟号	UH（mm）	UL（mm）	SH（mm）	SL（mm）
1				
2				
3				
4				
Σ				

2. 查放线图计算抗生素效价。
3. 在实际操作中，哪些因素对生物效价测定有影响？

51 Biological Assay of Antibiotics

Purpose

1. To be familiar with the basic principles of biological assay of antibiotics.
2. To determine the antibiotics concentration with the cylinder plate method.

Principle

The antibiotics concentrations are usually determined by microbial method which is based on the antimicrobial activity of antibiotics. We use inhibitory and bactericidal activity as assessment standard for antibiotic potency, and it is parallel to clinical use. The microbiological methods to determine the antibiotics concentration include: the dilution method, the turbidimetric method and agar diffusion method, of which the cylinder plate method is the most commonly used.

Cylinder plate method is based on the disk diffusion. Antibiotics concentration can be determined according to the inhibition zone diameter compared with standard sample. As a cylinder plate, four stainless tubes (inner diameter of 6. 0mm ± 0. 1mm, outer diameter of 8. 0mm ± 0. 1mm, height of 10. 0mm ± 0. 1mm) are placed on the agar plate previously inoculated with the sensitive bacteria. Then antibiotics sample and standard antibiotics solution are added into the tubes. The antibiotics in the tubes will diffuse radically and a clear zone or ring will present around an antibiotic disk incubated at 37℃ for 16 to 18h, which is named inhibition zone. The antibiotics concentration can be determined by comparing the diameter of inhibition zone between samples and standards because it is associated with the antibiotics concentration within the tube. The characteristics of cylinder plate method include its high sensitivity and exhibiting the antibiotic activity directly, so it is included in the pharmacopoeia of many nations as an internationally common method.

The cylinder plate method to determine antibiotics concentration can be classified as one – dosage method, two – dosage method and three – dosage method. The two – dosage method is the most commonly used one, and is also called four dots method. Standard antibiotics and samples are diluted to the concentration ratio of 2 : 1 or 4 : 1 and compare their antimicrobial activity with antibiotics sample on the same plate. The antibiotics concentration is calculated according to the linear relationship between logarithm amount of antibiotic concentration and the inhibition zone diameter. In this experiment, four stainless tubes are placed on double layer agar plate containing test bacteria in upper layer, and two tubes for high concentration and the other for low concentration.

After the determination of the inhibition zone diameter, we can calculate the antibiotics concentration as follows:

(1) Calculate W and V:

$$W = (SH + UH) - (SL + UL)$$

$$V = (UH + UL) - (SH + SL)$$

Where UH and UL are the inhibitory diameters of high and low concentration of antibiotics sample, SH and SL are the inhibitory diameters of high and low concentration of standard antibiotics.

(2) Calculate θ:

$$\theta = D \times \text{antilg}\ (IV/W)$$

Where θ is the concentration ratio of antibiotics sample and standard antibiotics, D is the high concentration ratio of antibiotics sample and standard antibiotics which is usually one, I is the logarithm of high and low concentration ratio which is usually lg2 or lg4.

(3) Calculate Pr:

$$Pr = Ar \times \theta$$

Where Pr is the determined concentration of antibiotics sample and Ar is the assumed initial concentration.

The concentration of antibiotics sample is also available from figure of W, V and θ as shown in Fig. 51 – 1.

The antibiotics standard, antibiotics sample, culture medium, culture condition and concentration of antibiotics are strictly required in the pharmacopoeia when determining antibiotics concentration. This experiment refers to determination of tetracyclines concentration in 2005 edition of pharmacopoeia.

Materials and Apparatus

1. Cultures

Micrococcus luteus [CMCC (B) 28001].

2. Media

Recipe (g/L) peptone: 10, beef extract: 6, glucose: 1, agar: 15 ~ 20, sterilize at 115℃ for 30 min.

3. Antibiotics

Tetracycline sample and standard (the ratio of high dosage and low dosage is 2 : 1, concentration range of tetracycline is 10.0 ~ 40.0 U/mg), sterile phosphate buffer pH 6. 0.

4. Reagents and apparatus

Sterilized physiological saline, sterile plate, Oxford cup (inner diameter of 6. 0mm ±0. 1mm, outer diameter of 8. 0mm ± 0. 1mm, height of 10. 0mm ± 0. 1 mm, sterile clay cover, sterile pipette, tweezers, dropper, ruler and so on.

Procedures

1. Prepare test bacteria suspension

Inoculate *Micrococcus luteus* [CMCC (B) 28001] to slant agar, then incubate at 26 ~ 27℃ for 24h. After that, rinse the lawn with 0.9% sterile NaCl solution and prepare the test bacteria suspension.

2. Prepare antibiotics solution

Precisely weigh a certain amount of the tetracyclines standard and sample according to the pharmacopoeia, dissolve with sterile phosphate buffer solution, dilute the ratio of high dosage and low dosage is 2 : 1, label with "SH" "SL" and "UH" "UL".

3. Prepare two layers plates

Take out four sterile plates whose diameter is 90 mm and height is 16 ~ 17mm, add 20ml sterile nutrient agar medium respectively to make it as bottom layer after cool. Dilute the spore suspension of *Micrococcus luteus* with sterilized physiological saline to a appropriate concentration in order to get an inhibition zone diameter of 18 to 22 mm for standard antibiotics solution. Absorb 1 ml of spore suspension to 200ml upper layer nutrient agar incubated at 50℃ and mix. Absorb 5ml rapidly into the plate above and cool.

4. Determination

Label SH, SL, UH and UL on the back of the plate. Sterilize tweezers on fire 3 times and pick up 4 sterile Oxford cups on the plate. (Note that don't let the cups into the media and keep the same distance between the cups.) Add drug solutions into 4 cups respectively, don't let them spill over to keep the same height of solutions within the cups, then cover the plate with the clay cover and stay it for 30 min.

5. Incubate the plate at 37℃ for 16 to 18h.

6. Observe the results

Replace the clay cover with a glass one, take out the Oxford cups and then measure the diameter of each inhibition zone (mm) with a ruler (SH、SL、UH、UL). The error among the four plates shoule be no more than 0. 1mm and then calculate the potency according to the formula.

Results and Discussion

1. Record diameter of inhibition zone and calculate antibiotic activity (standard tetracyclines: 1000U/mg) .

Table 51 - 1 Results of the inhibition zone

NO	UH (mm)	UL (mm)	SH (mm)	SL (mm)
1				
2				
3				
4				
Σ				

2. Calculate antibiotic activity according to the Fig. 51 - 1.

3. What are the factors affecting biological activity in practical operation?

实验五十二　药物无菌检查法

【实验目的】

熟悉药物制剂的无菌检查方法。

【实验原理】

无菌检查法是用于检查药典及法规要求无菌的药品、医疗器具、原料、辅料及其他品种是否无菌的一种方法，药典规定各种注射剂（如针剂、输液等）、手术眼科制剂等都必须保证无菌。无菌检查的全部过程应严格遵守无菌操作原则，以防止微生物的污染，影响结果的判定。无菌检查是用部分样品的测定结果推断整体的含菌情况，适当的试验操作技术能确保结果的科学、准确。

药物的无菌检查法包括直接接种法和膜过滤法。药典规定只要供试品性状允许，应采用膜过滤法。无菌检查用的培养基应符合培养基的灵敏度检查的要求，即加阳性菌的培养基管均生长良好。而当建立药品的无菌检查法时，也应进行方法的验证，选用合适的阳性对照菌进行验证，以证明所采用的培养基和检验方法适合于该药品的无菌检查。这些验证可在供试品的无菌检查前或与供试品的无菌检查同时进行。做验证试验时，在加入被测样品之后，将一定量阳性对照菌（一般不多于100CFU）加在最后冲洗滤膜的稀释液中（膜过滤法）或检测培养基中（直接接种法），培养5天。培养基的灵敏度检查和方法的验证选用的阳性试验菌如下：

金黄色葡萄球菌（*Staphylococcus aureus*）〔CMCC（B）26003〕

铜绿假单胞菌（*Pseudomonas aeruginosa*）〔CMCC（B）10104〕

枯草芽孢杆菌（*Bacillus subtilis*）［CMCC（B）63501］

生孢梭菌（*Clostridium sporogenes*）〔CMCC（B）64941〕

白色念珠菌（*Candida albicans*）〔CMCC（F）98001〕

黑曲霉（*Aspergillus niger*）〔CMCC（F）98003〕

方法验证试验中，与对照管比较，如含供试品各容器中的试验菌均生长良好，则说明供试品的该检验量在该检验条件下无抑菌作用或其抑菌作用可以忽略不计，可照此检查方法和检查条件进行供试品的无菌检查。如含供试品的任一容器中的试验菌生长微弱、缓慢或不生长，则说明供试品的该检验量在该检验条件下有抑菌作用，可采用增加冲洗量、或增加培养基的用量、或使用中和剂或灭活剂（如β－内酰胺酶、对氨基苯甲酸）、或更换滤膜品种等方法，消除供试品的抑菌作用，并重新进行方法验证试验。

在药典中，针对检验数量、检验量及不同性质的样品阳性对照菌株的选择，都有严格规定。参照药典，无抑菌作用及抗革兰阳性菌为主的供试品，以金黄色葡萄球菌为对照菌；抗革兰阴性菌为主的供试品，以大肠埃希菌为对照菌；抗厌氧菌为主的供试品，以生孢梭菌为对照菌；抗真菌为主的供试品，以白色念珠菌为对照菌。

【实验材料】

1. 待检药物

0.9%氯化钠注射液，注射用盐酸四环素。

2. 培养基

需气菌、厌气菌培养基（硫乙醇酸盐液体培养基），真菌培养基（改良马丁培养基），培养基配方见附录。

3. 试剂与器材

灭菌生理盐水，试管，灭菌吸管等。

4. 对照用试验菌

（1）金黄色葡萄球菌［*Staphylococcus aureus*，CMCC（B）26003］菌液　取金黄色葡萄球菌的营养琼脂斜面新鲜培养物1白金耳环，接种至需气菌、厌气菌培养基内，30～35℃培养16～18小时后，用灭菌生理盐水稀释成10^{-6}。

（2）生孢梭菌［*Clostridium sporogenes*，CMCC（B）64941］菌液　取生孢梭菌的需气菌、厌气菌培养基新鲜培养物1白金耳环，再接种至相同培养基内，30～35℃培养18～24小时后，用灭菌生理盐水稀释成10^{-5}。

（3）白色念珠菌［*Candida albicans*，CMCC（F）98001］菌液　取白色念珠菌的霉菌琼脂培养基斜面新鲜培养物1白金耳环，接种至霉菌培养基内，23～28℃培养24小时后，用灭菌生理盐水稀释成10^{-5}。

【实验方法】

（一）膜过滤法

本实验采用膜过滤法对注射用盐酸四环素作无菌检查。参照药典选用金黄色葡萄球菌为阳性对照菌。

膜过滤法应优先采用封闭式薄膜过滤器（见图52－1），也可使用一般的薄膜过滤器。无菌检查用的滤膜孔径应不大于0.45μm，直径约为50mm。根据供试品及其溶剂的特性选择滤膜材质。滤器及滤膜使用前应采用适宜方法灭菌。

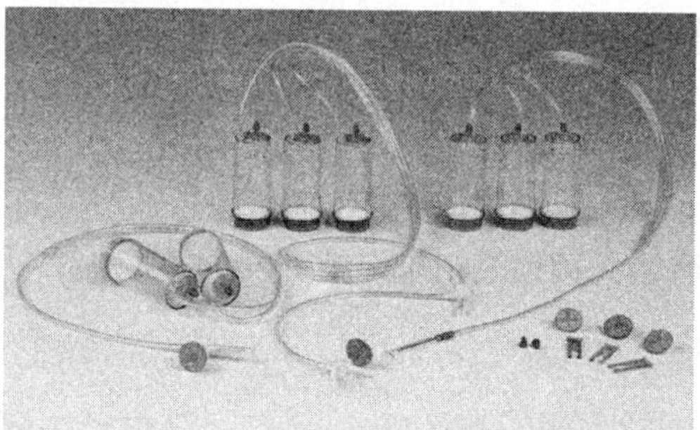

图52－1　全封闭集菌培养器

Fig. 52－1　The appara tus of membrane filtration

1. 准备培养基及供试液

配制3管硫乙醇酸盐液体培养基和2管真菌培养基。参照药典规定，样品用0.9%灭菌氯化钠溶液500ml溶解制成供试液。

2. 薄膜过滤及接种

先用少量稀释剂润湿滤膜；摇匀样品，以无菌操作将供试液加入薄膜过滤器内，减压抽干；同法用100 ml 0.9%灭菌氯化钠溶液冲洗滤膜，重复3次，冲洗清除残留在滤筒、滤膜上的抗生素；直接将各种检验用培养基50ml加入封闭的滤筒内（封闭式薄膜过滤器）；或取出滤膜，分成3等片（一般的薄膜过滤器），取2片滤膜分别放在2管均装50ml硫乙醇酸盐培养基中，其中1管接种对照用菌液1ml，供作阳性对照。另1片滤膜放在含50ml真菌培养基管中。另取1支硫乙醇酸盐培养基和1支真菌培养基作阴性对照。

3. 培养及观察

上述含培养基的容器按规定的温度培养14天。硫乙醇酸盐培养基培养温度30~35℃，改良马丁培养基培养温度23~28℃。培养期间应逐日观察并记录是否有菌生长。结果记入表52-1。

4. 结果判定

培养基中接种的阳性对照菌株培养3~5天应生长良好，阴性对照管培养基14天后应澄清无菌生长；而供试品管均澄清，或虽显浑浊但经确证无菌生长，判供试品符合规定；若供试品管中任何一管显浑浊并确证有菌生长，判供试品不符合规定。

（二）直接接种法

本实验采用直接接种法对0.9%氯化钠注射液作无菌检查。参照药典选用金黄色葡萄球菌为阳性对照菌。

1. 以无菌操作吸取1ml对照菌液或待测样品按表52-2所示加入含15ml的无菌试验培养基中，摇匀。

2. 需气菌、厌气菌培养基于30~35℃培养14日，霉菌培养基于23~28℃培养14日，阳性菌对照管培养1天，按表52-2所示记录实验结果。

3. 培养期间应逐日检查是否有菌生长，阳性对照24小时内应有细菌或真菌生长。

4. 结果判断

当阳性对照管显浑浊并确有细菌或真菌生长，阴性对照管无菌生长，药物试验的需气菌、厌气菌及霉菌培养基管均为澄清或显浑浊，但经显微镜检证明无菌生长，则判定被检测样品无菌试验合格。

【结果与讨论】

1. 将注射用盐酸四环素无菌检查结果记录于表52-1中。

表 52-1　注射用盐酸四环素无菌试验结果（培养基分装量 50ml，阳性对照菌接种量 1ml）

管号	样品	接种	培养时间（天）	需气菌、厌气菌	真菌
1	需气菌、厌气菌培养基	金黄色葡萄球菌 + 滤膜 1 片	3~5		
2	需气菌、厌气菌培养基	阴性对照	14		
3	需气菌、厌气菌培养基	滤膜 1 片	14		
4	真菌培养基	阴性对照	14		
5	真菌培养基	滤膜 1 片	14		

* 记录结果时“+”为浑浊，“-”为澄清。

2. 将 0.9% 氯化钠注射液无菌检查结果记录于表 52-2 中。

表 52-2　0.9% 氯化钠注射液无菌试验结果（培养基分装量 15ml，阳性菌接种量 1ml）

管号	样品	接种	培养时间（天）	需气菌、厌气菌	真菌
1	需气菌、厌气菌培养基	金黄色葡萄球菌	1		
2	需气菌、厌气菌培养基	生孢梭菌	1		
3	需气菌、厌气菌培养基	阴性对照	14		
4	需气菌、厌气菌培养基	0.9% N.S	14		
5	需气菌、厌气菌培养基	0.9% N.S	14		
6	需气菌、厌气菌培养基	0.9% N.S	14		
7	需气菌、厌气菌培养基	0.9% N.S	14		
8	需气菌、厌气菌培养基	0.9% N.S	14		
9	真菌培养基	白色念珠菌	1		
10	真菌培养基	阴性对照	14		
11	真菌培养基	0.9% N.S	14		
12	真菌培养基	0.9% N.S	14		
13	真菌培养基	0.9% N.S	14		
14	真菌培养基	0.9% N.S	14		
15	真菌培养基	0.9% N.S	14		

3. 哪些药物需无菌检查？抗生素类药物如何进行无菌检查？

52 Sterility Test of Drugs

Purpose

To be familiar with the test for sterility of pharmaceutical preparations.

Principle

Test for sterility is needed for pharmaceutical preparations, their raw materials and additives, some medical appliances and so on required in the pharmacopoeia are sterile. No microorganisms should be contained in the preparations for injection and eye in order to accord with pharmacopoeia. The whole process of sterility testing should be carried out under strict conditions specifically designed to avoid microbial contamination of the material being tested which can affect the determination of the results. Once a sample has been tested and is found to be free of contamination, the same degree of purity should apply to the whole batch. Adequate performance of the testing technique should be ensured periodically.

The test for sterility may be carried out by membrane filtration or by direct inoculation. The technique of membrane filtration is to be preferred whenever the nature of the product permits. The culture media used for the test for sterility meet the requirements of the growth promotion test of aerobes, anaerobes, and fungi to validate the effectiveness of the medium in the presence of the preparation to be tested. The media are suitable if a clearly visible growth of the microorganisms occurs. Validation test for the test for sterility of the product to be examined should be carried out before, or in parallel, with the test on the product to be examined using exactly the same methods, except for the following modifications. After transferring the content of the container or containers to be tested, add an inoculum of a small number of viable microorganisms (not more than 100 cfu) to the final portion of sterile diluent used to rinse the filter (Membrane Filtration) or to the medium (Direct Inoculation). Incubate all the containers containing medium for not more than 5days. Strains of the test microorganisms suitable for use in the growth promotion test and the validation test are as follows: *Staphylococcus aureus* [CMCC (B) 26003], *Pseudomonas aeruginosa* [CMCC (B) 10104], *Bacillus subtilis* [CMCC (B) 63501], *Clostridium sporogenes* [CMCC (B) 64941], *Can-*

dida albicans [CMCC (F) 98001], *Aspergillus niger* [CMCC (F) 98003].

If clearly visible growth of microorganisms is obtained after the incubation, visually comparable to that in the control vessel without product, either the product possesses no antimicrobial activity under the conditions of the test or such activity has been satisfactorily eliminated. The test for sterility may then be carried out without further modification. If clearly visible growth is not obtained in the presence of the product to be tested, visually comparable to that in the control vessels without product, the product possesses antimicrobial activity that has not been satisfactorily eliminated under the conditions of the test. Modify the conditions in order to eliminate the antimicrobial activity, and repeat the validation test.

Materials and Apparatus

1. Drug

0.9% physiological saline injection solution, Tetracycline Hydrochloride for injection.

2. Culture media

Aerobic bacteria and anaerobic bacteria culture media (fluid thioglycollate medium), fungi culture media (modified Martin culture media).

3. Reagent and apparatus

Sterile physiological saline, test tubes and sterile sucker.

4. The control test bacteria

(1) *Staphylococcus aureus*, CMCC (B) 26003 Inoculate a full loop of the nutrient agar slant culture of *Staphylococcus aureus* to a fluid thioglycollate medium, incubate at 30 to 35℃ for 16 to 18h, dilute with sterilized physiological saline to 10^{-6}.

(2) *Clostridium sporogenes*, CMCC (B) 64941 Inoculate a full loop of the nutrient agar slant culture of *Clostridium sporogenes* to a fluid thioglycollate medium, incubate at 30 to 35℃ for 18 to 24h, dilute with sterilized physiological saline to 10^{-5}.

(3) *Candida albicans*, CMCC (F) 98001 Inoculate a full loop of the nutrient agar slant culture of *Candida albicans* to a fungi media, incubate at 23 to 28℃ for 24h, dilute with sterilized physiological saline to 10^{-5}.

Procedures

A. Membrane Filtration

In this experiment, the sample is Tetracycline Hydrochloride for injection, using *Staphylococcus aureus* as a positive control.

Use membrane filters having a nominal pore size not greater than 0.45 m and about

50mm in diameter whose effectiveness to retain microorganisms has been established. Cellulose nitrate filters, for example, are used for aqueous, oily, and weakly alcoholic solutions; and cellulose acetate filters, for example, are used for strongly alcoholic solutions. Specially adapted filters may be needed for certain products (*e. g.*, for antibiotics). The filtration apparatus and membrane are sterilized by appropriate means. The apparatus is designed so that the solution to be examined can be introduced and filtered under aseptic conditions: it permits the aseptic removal of the membrane for transfer to the medium, or it is suitable for carrying out the incubation after adding the medium to the apparatus itself.

1. Preparation of the culture media and the drug solution

Prepare 50ml fluid thioglycollate media in three test tubes or flasks and 50ml fungi culture media in two test tubes or flasks respectively. Dissolve the Tetracycline Hydrochloride for injection with 500 ml of sterile 0.9% physiological saline to prepare the drug solution as sample.

2. Filtration and innoculation

Transfer a small quantity of a suitable, sterile diluent onto the membrane in the apparatus and filter. Shake the sample solution uniformly, add it to the Membrane Filter under aseptic techniques. If the product has antimicrobial properties, wash the membrane not less than three times by filtering through it each time the volume of the chosen sterile diluent used in the validation test. Do not exceed a washing cycle of 5 times 200ml, even if during validation it has been demonstrated that such a cycle does not fully eliminate the antimicrobial activity. Transfer the culture media to the filtration apparatus or cut the whole membrane aseptically into 3 equal parts, and transfer 2 of them into the 2 tubes of 50ml fluid thioglycollate media respectively, and inoculate 1ml control bacteria solution to one of them as positive control. Put another piece of membrane to 1 tube of fungi culture medium. Choose the last tube of 50 ml fluid thioglycollate medium and the last tube of fungi culture medium as negative control.

3. Incubate and observe

Incubate the above containers with culture media at stated temperature for 14 days. (Maintain the temperature at 30 ~ 35℃ for media mainly intended to detect bacteria and at 23 ~ 28℃ for media specifically intended to detect fungi.) During the period of incubating, observe and record whether there is microorganism every day. Record the results as shown in Table 52 – 1.

4. Results Assessment

The media thus inoculated with the reference strains are satisfactory if clear evidence of growth appears in all inoculated media within 3 ~ 5 days and the negative control media is no evidence of growth appears after 14 days. If no evidence of microbial growth is found, the product to be examined complies with the test for sterility. If evidence of microbial growth is found, the product to be examined does not comply with the test for sterility.

B. Direct Inoculation

In this experiment, the sample is the 0.9% physiological saline injection solution, using *Staphylococcus aureus* as a positive control.

1. Absorb 1ml bacteria solutions diluted above to a tested tube containing fluid thioglycollate medium and fungi media shown as Table 52 – 2 and mix it gently.

2. Incubate fluid thioglycollate medium at 30 to 35℃ for 14 days and fungi media at 23 to 28℃ for 14 days except one day for controls. Record the results as shown in Table 52 – 2.

3. Check whether there is microbial growth every day. Controls should be observed for microbial growth within one day.

4. Results judgement

When positive and negative control tubes are as usual and tested tubes show no growth, it means that the drug test is qualified for aseptic inspection.

Results and Discussion

1. Record the results of Membrane Filtration in Table 52 – 1.
2. Record the results of Direct Inoculation in Table 52 – 2.
3. What kinds of drugs need asepticiam? How to perform asepticism of antibiotics?

实验五十三　微生物限度检查法

【实验目的】

1. 熟悉平板菌落计数法测定药物中细菌、霉菌和酵母菌的总数，检查药物是否符合微生物限度标准。

2. 了解不同药物的微生物限度标准。

3. 了解药物中控制菌的检查方法。

【实验原理】

口服药及外用药，在生产和临床使用过程中，不要求达到完全无菌，但是为了保证药品的质量，防止药品的污染，需要限制性控制微生物的数量和种类。目前对口服药及外用药物的微生物学检验主要是微生物限量检验，我国药典规定的检查项目包括：细菌总数测定、霉菌和酵母菌总数测定；控制菌的检验其中包括：大肠埃希菌、沙门菌、金黄色葡萄球菌、铜绿假单胞菌、破伤风梭菌等病原菌检查和活螨的检验等。细菌和霉菌总数的测定采用平皿法和膜过滤法；根据细菌的形态结构和生理生化特性来检查病原菌的存在；活螨可用显微镜检法。

1. 细菌总数的测定是检查每克或每毫升被检药品内所含有的活菌数（实际是需氧菌的活菌数），以判断供试药物被细菌污染的程度；霉菌和酵母菌总数的测定是考察供试药物中每克或每毫升所含的活的霉菌和酵母菌的总数，以判明供试药物被真菌污染的程度；是对该药物生产过程卫生学总评价的一个重要依据。

细菌和霉菌的平皿法和膜过滤法计数应按规定报告结果。平皿法中细菌宜选取平均菌落数在 30 ~ 300 之间的稀释级，霉菌宜选取平均菌落数在 30 ~ 100 之间的稀释级；而膜过滤法的每片滤膜上的菌落数应不超过 100 个；作为报告菌落计数的依据。其他情况按下列方法报告结果：

（1）如有 1 个稀释级在 30 ~ 300（30 ~ 100）之间时，将该稀释级的菌落数乘以稀释倍数报告。

（2）如同时有 2 个稀释级在 30 ~ 300（30 ~ 100）之间时，按下式计算两级比值 R。式中：N_H、N_L分别为高低稀释级的菌落数，D_H、D_L分别为高低稀释级的稀释倍数。当 R ≤2 时，以两稀释级的菌落数乘以稀释倍数的均值报告；当 2 < R≤5 时，以低稀释级平均菌落数乘以稀释倍数报告；当 R > 5 时，或高稀释级的菌落数大于或等于低稀释级的菌落数等异常情况时，应查明原因再行检查，必要时，应进行方法的重新验证。

$$R = \frac{N_H \times D_H}{N_L \times D_L}$$

（3）如各稀释级的平均菌落数均小于 30 时，一般按最低稀释级的菌落数乘以稀释倍数报告。

（4）如各稀释级的平板均无菌落生长，或仅最低稀释级的平板有菌落生长，但平均菌落数小于1时，以小于1乘以最低稀释倍数的值报告菌数。

2. 控制菌的测定可根据细菌的形态结构和生理生化特性来检查，一般的检验程序如下：

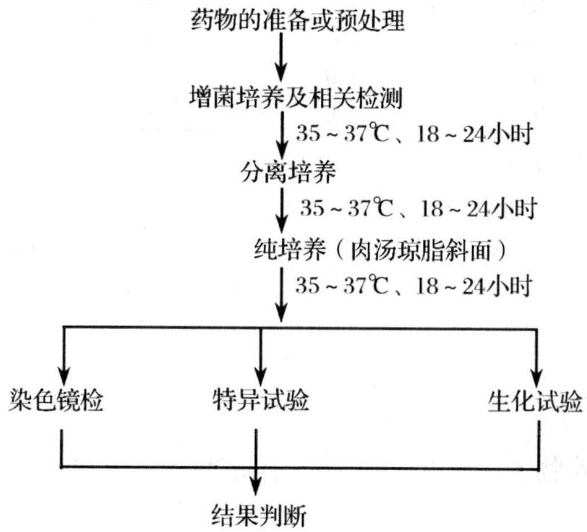

本实验只选作其中之一，即大肠埃希菌的检查。大肠埃希菌是口服药品的常规必检项目之一。大肠埃希菌是人和温血动物肠道内寄生的正常菌群，药品中的大肠埃希菌来源于人和温血动物的粪便。凡由供试品中检出大肠埃希菌时，表明该药品可能已被粪便污染，也就可能被存在于粪便的其他肠道致病菌所污染。患者服用这种药物后，就可能出现这些病原体感染的危险。因此，口服药品中不得检出大肠埃希菌。

药物的微生物限度标准如表53－1所示。含动物组织来源的制剂，不得检出沙门菌；口服药品中不得检出大肠埃希菌；凡外用约和眼科制剂不得检出金黄色葡萄球菌和铜绿假单胞菌。抗细菌的口服抗生素制剂应检查霉菌，每克中不得超过100个；抗真菌的口服抗生素制剂应检查细菌，每克不得超过100个；霉变、长螨者均为不合格药品。

表53－1　微生物限度标准（个/g或个/ml）*

剂型	细菌数	霉菌、酵母菌数	大肠埃希菌	金黄色葡萄球菌	铜绿假单胞菌
片剂	1000	100	—		
酊剂	100	100	—		
栓剂	100	10		—	
胶囊剂	1000	100	—		
软膏剂	100	100		—	
一般眼膏剂	100	—		—	
一般滴眼剂	100			—	
丸剂（滴丸、糖丸）	1000	100	—		
气雾剂	100	10	—	—	—
糖浆剂	100	100	—		
膜剂	100/10cm	10/10cm^2	—	—	—
颗粒剂	1000	100	—		
口服溶液剂、混悬剂、乳剂	100	100	—		

续表

剂型	细菌数	霉菌、酵母菌数	大肠埃希菌	金黄色葡萄球菌	铜绿假单胞菌
散剂	1000	100	—		
外用散剂	100	100		—	—
滴耳剂	100	10	—	—	—
滴鼻剂	100	10	—	—	—
洗剂	100	100	—	—	—
搽剂	100	100	—	—	—
凝胶剂	100	100	—	—	—

* "—"为每1g或1ml中不得检出。

【实验材料】

1. 药物

咳嗽糖浆。

2. 对照菌

大肠埃希菌液体培养物。

3. 培养基

营养琼脂，玫瑰红钠琼脂，酵母浸出粉胨葡萄糖琼脂（YPD），胆盐乳糖培养基（BL），曙红亚甲蓝琼脂培养基（EMB）等。

4. 试剂及器材

灭菌生理盐水，无菌试管，无菌吸管，无菌平皿等。

【实验方法】

（一）细菌、霉菌和酵母菌总数测定

1. 药物配制

将待测咳嗽糖浆摇匀，用吸管吸取10ml并加入到90ml灭菌生理盐水中作为供试液，取1ml供试液，以9ml灭菌生理盐水进一步稀释成1∶10、1∶100、1∶1000的浓度。

2. 接种

（1）倾注倒平板法　分别吸取上述稀释的溶液各1ml于无菌平皿中，再加入15ml恒温于45℃的营养琼脂培养基（细菌计数）、玫瑰红钠琼脂培养基（霉菌计数）、酵母浸出粉胨葡萄糖琼脂培养基（酵母菌计数），混匀，待凝固后将平板于30~35℃或23~28℃倒置培养48小时。每个稀释度取3个平皿作平行试验。

（2）薄膜过滤法（有抑菌作用的样品用此法）　采用无菌薄膜，滤膜孔径应不大于0.45μm，直径约为50mm，将供试液过滤后，再将滤膜置于培养基上进行菌落计数。一个滤膜对应接种于一个平板，每个样品平行做3个平板。

3. 计数菌落，写报告

计算各稀释级的平均菌落数，按规则报告细菌、霉菌和酵母菌含量。

（二）大肠埃希菌的检查

1. 增菌培养

准备3瓶内装100ml胆盐乳糖培养基的三角瓶。将待测咳嗽糖浆摇匀，分别取10ml加入到2份中，其中1份再加入对照菌液1ml作阳性对照，第3份100ml胆盐乳糖培养基中加入与供试液等量的稀释剂作空白对照。于35～37℃恒温培养18～24小时（必要可延至48小时）。空白对照瓶应无菌生长。其他2瓶培养液变浑浊表明有细菌生长。

2. MUG及靛基质检测试验

取上述培养物0.2 ml接种至含5 ml MUG培养基的试管内，培养，于5小时、24小时在366nm紫外线下观察，同时用未接菌的MUG培养基作本底对照。若管内培养基呈现荧光，为MUG阳性；不呈现荧光，为MUG阴性。观察后，沿培养基的管壁加入数滴靛基质试液，液面呈玫瑰红色，为靛基质阳性；呈试剂本色，为靛基质阴性。本底对照应为MUG阴性和靛基质阴性。

如MUG阳性、靛基质阳性，判供试品检出大肠埃希菌；如MUG阴性、靛基质阴性，判供试品未检出大肠埃希菌；如MUG阳性、靛基质阴性，或MUG阴性、靛基质阳性，则应继续下述检查。

3. 平板分离培养

将上述培养物对应划线接种于曙红亚甲蓝琼脂平板或麦康凯琼脂平版，于35～37℃恒温培养18～24小时，当阳性对照的平板呈现阳性菌落时，供试品的平板无菌落生长，或有菌落但不同于表53－2所列的特征，可判为未检出大肠埃希菌。

表53－2　大肠埃希菌菌落形态特征

培养基	菌落形态
曙红亚甲蓝琼脂	呈紫黑色、浅紫色、蓝紫色或粉红色，菌落中心深紫色或无明显暗色中心，圆形，稍凸起，边缘整齐，表面光滑，湿润，常有金属光泽
麦康凯琼脂	鲜桃红色或微红色，菌落中心深桃红色，圆型，扁平，边缘整齐，表面光滑，湿润

如生长菌落与表53－2所列特征相符或疑似者，应挑选2～3个菌落分别接种于营养琼脂培养基斜面，培养18小时，作以下检查。

4. 显微镜检

取上述斜面培养物，革兰染色后显微镜检，大肠埃希菌为革兰阴性无芽孢的短杆菌。

5. 生理生化反应

取上述斜面培养物，分别接种于乳糖发酵管，均进行IMViC试验，培养24～48小时观察。

大肠埃希菌可发酵乳糖产酸产气。

大肠埃希菌的IMViC试验结果为“＋　＋　－　－”（IMViC分别代表：靛基质试验、甲基红试验、乙酰甲基甲醇生成试验及枸橼酸盐利用试验）。

当空白对照试验呈阴性，供试品检查为革兰阴性无芽孢杆菌；乳糖发酵产酸产气或产酸不产气；IMViC试验为阳性、阳性、阴性、阴性或阴性、阳性、阴性、阴性，判为检出大肠埃希菌。对可疑反应的菌株，应重新分离培养后，再作生化试验证实。

6. 观察结果，给出检查报告

咳嗽糖浆未检出大肠埃希菌时为控制菌检验合格。

【结果与讨论】

1. 判断咳嗽糖浆的细菌、霉菌、酵母菌总数及大肠埃希菌检查结果，是否符合药物的微生物限度标准。

2. 如何判断某一药物中控制菌符合微生物限度标准？举例说明。

附：4－甲基伞形酮葡糖苷酸（MUG）培养基配方

胨	10.0 g
硫酸锰	0.5 mg
硫酸锌	0.5 mg
硫酸镁	0.1 g
氯化钠	5.0 g
氯化钙	50 mg
磷酸二氢钾	0.9 g
磷酸氢二钠	6.2 g
亚硫酸钠	40 mg
去氧胆酸钠	1.0 g
MUG	75 mg
水	1000 ml

除 MUG 外，取上述成分混合，微温溶解，调节 pH 值使灭菌后为 7.3 ± 0.1，加入 MUG，溶解，每管分装 5 ml，灭菌。

53 Microbial Limit Test

Purpose

1. To be familiar with determining the number of bacteria, molds and yeast in drug and check whether the drug accord with microbial standards in pharmacopoeia.

2. To understand the microbial limit standards of different drugs.

3. To learn how to check the controlled bacteria in drug.

Principle

In the process of manufacture and clinical use, drugs for oral or external need microbial limitation check except for aseptic inspection in order to guarantee their quality and prevent them from contamination. That is the number of bacteria and molds and yeast, the pathogenic bacteria (or the limited bacteria) such as *Escherichia coli*, *Salmonella species*, *Staphylococcus aureus*, *Pseudomonas aeruginosa*, *Bacillus tetani* and living mite according to the pharmacopoeia. The number of bacteria and molds can be determined by Pour plate method and Membrane Filtration. Pathogenic bacteria can be confirmed by its morphology biochemical tests while living mite is observed by microscope.

1. The determination of total bacterial number is to determine the quantity of viable bacteria per gram or milliliter in the sample (actually the number of viable number of aerobic bacteria) in order to assess the extent of bacterial contamination. The determination of total number of molds and yeasts is to detect the total number of viable molds and yeasts per gram or milliliter, then further judge the extent of contamination in the sample, it is the most important basis of hygienic assessment in the process of drug manufacture.

The number of bacteria and molds in the plates is reported in the following way. At first, it is suitable for bacteria counting when there are 30 to 300 colonies in plates while 30 to 100 for molds. The number of bacteria and molds in each membrane is not more than 100.

(1) When the number of colonies at one dilution series is among 30 to 300 for bacteria or 30 to 100 for molds, report it.

(2) When the number of colonies at two dilution series is among 30 to 300 for bacteria or 30 to 100 for molds, calculate the ratio as follows:

Where N_H and N_L are colonies of high and low dilution series respectively, D_H、D_L are dilution times for two dilution series respectively. When $R \leqslant 2$, the average number of two dilution series is reported and the number of low dilution series is reported for $2 < R \leqslant 5$. When $R > 5$ or the average number of high dilution serie is more than or equal to the average number of low dilution serie, the validation test should be carried out.

$$R = \frac{N_H \times D_H}{N_L \times D_L}$$

(3) When the average number of colonies at all dilution series is less than 30, report colonies number based on the lowest dilution series.

(4) When the average number of colonies at all dilution series is nothing or only the average number of colonies at the lowest dilution serie is less than 1, report colonies number based on the lowest dilution series.

2. The determination of the limited bacteria can be checked by morphous, structure and physiological characteristics of bacteria. The general checking procedures are as follows:

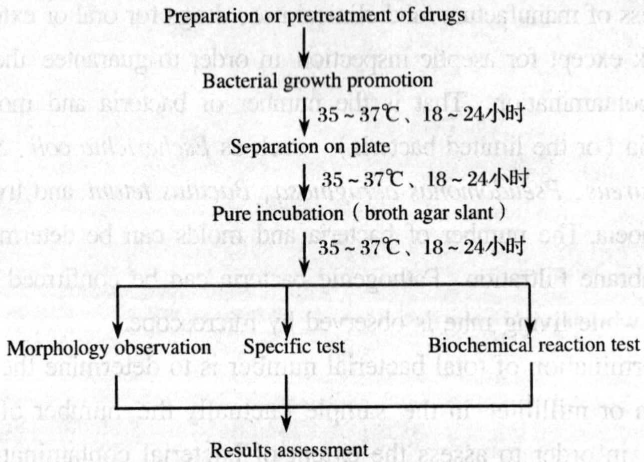

In this experiment, we mainly introduce the checking of *E. coli*. *E. coli* is one of the limited bacteria that must be checked for oral drugs. *E. coli* is the normal colony that parasited in the intestinal tract of human and warm - blooded animal, most of *E. coli* in drugs is from feces of human and warm - blooded animal. Once *E. coli* is detected from sample, it is to say the drugs may have been contaminated by feces, and it may also be contaminated by other intestinal tract pathogenic bacteria that exist in the feces. After taking drugs, the risk of pathogenic infection may occur. So no *E. coli* can be detected for oral drugs.

The microbial standards for drugs are as shown in table 53 - 1. In addition, no *Salmonella species* should be detected for preparations from any animal organs; no *E. coli* should be detected for oral drugs; no *S. aureus* or *Pseudomonas aeruginosa* should be detected for external or ophthalmologic preparation. The number of molds in oral antibiotics should be controlled to be less than 100 CFU/g and bacteria in antifungi orals less than 100 CFU/g too. On the other hand, mildewed and mite contaminated drugs are not qualified.

Table 53-1 Microbial limit standards (CFU/g or CFU/ml)

Preparations	No. of bacteria	No. of molds and yeast	*E. coli*	*S. aureus*	*P. aeruginosa*
Tablets	1000	100	—		
Tincture	100	100	—		
Suppository	100	10		—	—
Capsules	1000	100	—		
Ointments	100	100		—	—
Eye ointments	100	—		—	—
Eye drops	100	—		—	—
Pills	1000	100	—		
Aerosol	100	10	—	—	—
Syrups	100	100	—		
Films	$100/10cm^2$	$10/10cm^2$	—	—	—
Granules	1000	100	—		
Oral liquids					
Suspensions	100	100	—		
Emulsions					
Powders	1000	100	—		
Powders for external use	100	100		—	—
Ear drops	100	10		—	—
Nasal drops	100	10	—	—	—
Lotions	100	100		—	—
Liniments	100	100		—	—
Gel	100	100	—	—	—

• "—" no detection in 1g or 1ml.

Materials and Apparatus

1. Drug

cough syrup.

2. Cultures

Positive control bacterium—*E. coli* broth cultures.

3. Media

Nutrient agar, Rose Bengal agar, YPD, BL and EMB or MacConkey.

4. Reagents and apparatus

Sterilized physiological saline, sterile test tube, sterile sucker, sterile plate.

Procedures

A. The number determination of bacteria, molds and yeasts

1. Preparation of drug solution

Mix cough syrup and absorb 10ml with sterile sucker to 90ml sterilized physiological saline, and then 1ml of the above drug solution is dissolved in 9ml physiological saline to get a series of diluted solutions of 1 : 10, 1 : 100 and 1 : 1000.

2. The number determination of bacteria

(1) Pour plate method Absorb 1ml of the diluted drug solution above in three plates respectively and pour 15ml nutrient agar (The number determination of bacteria); 15ml Rose Bengal agar (The number determination of molds); 15ml YPD media (The number determination of yeasts) incubated at 45℃, mix gently and cool. Incubate the plates at 30 to 35℃ (or 23 to 28℃) for 48h (or 72h).

(2) Membrane Filtration Shake the sample solution uniformly, add it to the Membrane Filter with filter membrane whose diameter is 50mm and aperture is no more than 0.45μm ± 0.02μm under aseptic techniques. After vacuum drying, transfer the membrane to an agar plate (one membrane should be inocubated to one agar plate. Repeat 3 times for one sample.)

3. Counting the colonies and record the report

Calculate the average number of colonies for each dilution series, report the number of bacteria, molds and yeasts.

B. Assay of *Escherichia coli*

1. Bacterial growth promotion

Mix cough syrup gently, inoculate 10 ml to 100 ml BL media in two flasks respectively, and add 1 ml of *Escherichia coli* broth cultures to one of the two as positive - control, and add the equivalent amount of diluted solution (10ml) to 100ml BL media in the third flask as blank control. Incubate at 35 ~ 37℃ for 18 to 24h. No bacteria grow in blank control. The media would become turbid if the pharmaceutical preparation was contaminated by bacteria.

2. MUG and Indole test

Transfer 0.2 ml of the cultures above – mentioned to 5ml of MUG medium and incubation. Observe it under the 366nm UV at 5h and 24h and take the MUG medium as negative control. If the cultures in tube present fluor, it show positive of MUG test; otherwise it is negative. After observation, add several drops of Indole test solution to the tube, if the upper of solution show rose bengal, it is positive of Indole test; otherwise it is negative.

When MUG test and Indole test are all positive, report to check out *Escherichia coli*; while MUG test and Indole test are all negative, report that *Escherichia coli* is not checked out. When positive of MUG test and negative of Indole test, or negative of MUG test and positive of Indole test are shown, the post – mentioned produres should be done.

3. Separation on plate

Inoculate the cultures of BL on three EMB (or MacConkey) media plates respectively and incubate at 35 ~ 37℃ for 18 to 24h. There is no bacterial colony whose characteristics listed in table 53 – 2 growth on the sample plate if the positive control plate represents positive colony, we can report that there is no *Escherichia coli* detected.

Table 53 – 2 morphological characteristic of *E. coli*

Culture media	Morphology of colony
EMB agar	black purple, light purple, blue purple or pink, the center of colony is black purple or no apparent dark center, round shape, elevation, trim edge, smooth surface, moisture and metal luster
MacConkey agar	pink or red, the center of colony is pink, round shape, thin and flat, trim edge, smooth surface, moisture

If the colonies are in accordance with table 53 – 2, we should inoculate 2 – 3 colonies to nutrient agar slant, incubate for 18h and do the following check.

4. Assay with microscope

By the observation with microscope, *Escherichia coli* show a short bacillus of gram negative.

5. Biochemical reaction test

Inoculate the above slant cultures to lactose fermentation tubes and each of IMViC tubes, incubate for 24 ~ 48 hours.

Escherichia coli can ferment lactose and produce acid and air.

The IMViC tests of *Escherichia coli* are "+ + – –".

When the blank control is negative, the bacteria in the sample are checked as Gram negative, spore – free bacillus, the fermentation lactose test is to produce acid and air or only acid but no air and IMViC test is "+ + – –" or "– + – –", we can report that the *Escherichia coli* is detected. For suspected cultures, please repeat the biochemical test after separation incubation.

6. Observe the results and fill in the checking reports.

The cough syrup is qualified when *Escherichia coli* is not checked out.

Results and Discussion

1. According to the number of bacteria, molds and yeasts of cough syrup and the checking results of *Escherichia coli*, is it qualified?

2. How to judge if a pharmaceutical preparation is qualified for microbial contaminations or not? Would you please illustrate with examples?

实验五十四　药物对流感病毒抑制作用的研究

病毒是最小的生物病原体，许多疾病由病毒感染引起。由于抗病毒药物的研制起步较晚．目前在临床上能有效地治疗病毒性疾病的药物十分匮乏。抗病毒药物药效学研究的主要内容包括体外细胞培养和体内动物试验两大部分。体外部分主要通过观察病毒感染后细胞活性的改变和病毒指标（CPE、病毒空斑、病毒感染复数等）的变化，初步判断药物的药效。体内部分通过建立病毒感染的动物模型、检测病毒对动物死亡率、脏器病理变化、脏器中病毒量等的影响，更为有效地判断药物的抗病毒药效。

病毒病种类繁多，病原类型各有不同，每种病原对细胞、组织和动物的亲嗜性亦各不相同。这些都给抗病毒药物的研制带来重重困难。本实验介绍的是针对流感病毒的药物抗病毒药效学分析基本手段。针对其他病毒的药物研制尚需要根据情况具体分析。

【实验目的】

1. 熟悉体外抗流感病毒试验的原理。
2. 了解体外培养流感病毒的方法。
3. 掌握 CPE 观察法和血凝法检测流感病毒。
4. 熟悉检测抗病毒药物体外药效的方法

【实验原理】

流感病毒流感病毒属正黏病毒，传播快，易感染，由于流感病毒的抗原易变化，疫苗需重复接种。因此其预防和治疗流感成为重要课题。流感的体外感染以 MDCK 细胞（狗肾细胞）为感染对象，体内感染多以小鼠为模型。本实验主要介绍体外的抗病毒药物研究方法。与其他病毒相比，除了用 CPE 法检测病毒感染及药物的影响外，正黏病毒由于包膜中含有能够凝集红细胞的血凝素，检测可采用更为简便的血凝实验。

【材料与试剂】

1. 细胞

MDCK 细胞。

2. 试剂

DMEM 培养液，含 10% 胎牛血清的 DMEM 培养液，Hanks 溶液，0. 25% 胰酶和 DEAE 消化液，5. 6% 或 7. 5% 的 $NaHCO_3$。

3. 其他

25ml 细胞培养瓶，96 孔细胞培养板，血凝试验板，无菌滴管，血球计数板，酒精灯，超净工作台，酒精棉球。

【实验方法】

1. 待测药物对 MDCK 细胞的毒性作用

（1）MDCK 细胞按 5×10^4/ml 浓度接种 96 孔培养板，每孔 100μl，置 37℃　5% CO_2 培养箱中培养 24 小时。

（2）待 MDCK 细胞形成单层，将待测药物等比稀释成不同浓度，分别加入。每孔加量 100μl，每个浓度接种 4 个复孔。同时设正常细胞对照组。

（3）96 孔培养板置于 37℃、5% CO_2 中培养，连续 3 日逐日观察细胞病变情况，并判定供试品的最大无毒稀释度（TC0）。

2. 待测药物对流感病毒的抑制作用

（1）利用细胞毒性实验结果，以待测药物的 TC0 为最高浓度，用 DMEM 培养液将之倍比稀释得到 7 个梯度，每个梯度设 4 个复孔。

（2）向 MDCK 单层细胞中同时加入上述不同稀释度的待测药物，37℃孵育 1 小时，弃毒液，加入不同浓度待测药物 100μl，于 37℃、5% CO_2 中培养。

（3）实验同时设正常细胞组、病毒对照组。48 小时后观察细胞病变。

CPE 特征为：细胞肿胀圆化，细胞间隙增大，细胞核固缩或碎裂，严重时细胞部分或全部脱落。

同时收获细胞培养上清，进行后述的红细胞凝集活性检查。

3. 红细胞凝集活性（HA）检查

（1）用毛细吸管取维持液 3～4 滴，放入血凝板孔中。

（2）加入等容量的 1% 豚鼠红细胞，轻摇混之，置室温或 4℃，1 小时后观察结果。

4. 通过将实验组、细胞对照组、病毒对照组的 CPE 结果观察，红细胞凝集活性（HA）检查结果进行对照，初步判断待测药物对流感病毒的抑制作用。

【结果与讨论】

1. 记录 CPE、红细胞凝集实验的结果。

2. 如某种药物能够抑制流感病毒的吸附。体外试验中，何时加入药物，能够有效地检测其抑制能力强弱。

54 Drugs Inhibition to Influenza Virus

Virus is the smallest biological pathogens. And most infectious diseases caused by viral infection. The development of antiviral drugs is later. At present, the clinical effective treatment for viral diseases is very lack. Pharmacodynamic study of antiviral drugs includes two ways, *in vitro* cell culture and *in vivo* animal experiments the most. *In vitro*, observing changes in cell activity and virus indicators (CPE, virus plaque, the MOI, *etc.*) are used to judge of the efficacy of drugs. *In vivo*, through establishing animal model of virus infection, testing animal mortality, pathological changes in organs, viral load in organs, more effective of antiviral drugs efficiency should be determined.

A wide range of virus diseases, different types of pathogens, different tropism of each pathogen on the cell , tissue and animal, these make the development of antiviral drugs more difficult. Described in the experimental is the basic means for analysis antiviral effect of drug aim at influenza viruses. Drugs for other viruses still need to develop case – by – case analysis

Purpose

1. To be familiar with the principle of anti – influenza virus test.
2. To be realize how to culture influenza virus *in vitro*.
3. To be grasp detecting influenza virus by CPE and hemagglutination.
4. To be familiar with the methods to detect antiviral effect of drugs *in vitro*.

Principle

Influenza virus is belong to influenza virus genus of the Orthomyxoviridae. Because of spreading fast, susceptible to infection, and virus antigens are easy to change, vaccination need to repeat, the prevention and treatment of influenza has become an important issue. *In vitro*, MDCK cells (dog kidney cells) is the influenza virus infection objects. *In vivo* mice is the animal model. in this experiment antiviral research methods in vitro are introduced. Compared with other viruses, the envelope of influenza virus contains hemagglutinin. So, in addition to CPE assay, the HI test is easier to detect the influenza viral load.

Materials and Apparatus

1. Cell line

MDCK.

2. Reagent

20% SRBC, DMEM cell culture medium, DMEM Cell culture medium with 10% FCS, Hanks solution, 0.25% trypsin, 0.02% DEAE, 5.6% $NaHCO_3$.

3. Others

25ml cell culture bottles, 96 – hole cell culture plate, hemagglutination board, sterile dropper, blood cell count board, alcohol lamp, clean Benches, alcohol swab.

Procedures

1. Test the toxicity of drugs to MDCK cells

(1) MDCK cells are inoculated into 96 – hole cell culture plate with the concentration of 5×10^4/ml, 100μl per hole. Then culture in 37℃ 5% CO_2 for 24h.

(2) When MDCK cell form monolayer, discard supernatant, then add 100μl drug to each hole. The drug is diluted into different concentration for 4 parallel – hole. At the same time, set up the control group of normal cells.

(3) 96 – well plate placed in 37 ℃, 5% CO_2 for 3 days. Observing the cytopathic effect everyday and determine the maximum non – toxic dilution (TC0).

2. The inhibition of test drugs on influenza virus

(1) According to the results of cytotoxicity experiments, make the TC0 as the highest concentration and 2 – fold dilute with DMEM culture medium to seven gradient concentration. Each gradient setting four parallel – holes.

(2) Add the different dilution of the test drug to the MDCK monolayer cells at the same time, incubate in 37℃ for 1h. Then dispose supernatant and add different concentrations of test drug 100μl. Keep on culturing at 37℃, 5% CO_2.

(3) the normal cells group and the virus control group are set up at the same time. 48 hour later, the CPE will been observed.

CPE characterization: cell swelling, cell gap increased, nuclear pyknosis or fragmentation, some or all cells fall off if seriously.

Cell culture supernatant harvested at the same time to check the red blood cell agglutination activity.

3. Erythrocyte agglutination (HA) activity

(1) Drip 3 ~ 4 drops of maintenance liquid into the hole of hemagglutination plate with capillary

(2) Adding 1% guinea pig red blood cell with equal content. Mixed gently, then

culture at room temperature or 4 ℃ fo 1h. Then observe the results.

4. To determine the inhibition of the drugs on influenza virus by analyse the results of CPE and hemagglutination activity (HA) test.

Results and Discussion

1. Record the results of CPE and hemagglutination.

2. If a drug could inhibit the adsorption of influenza virus. *In vitro* experiments, when adding the drug, which can effectively detect the strength of its activity?

附　　录

Ⅰ. 常用无菌玻璃仪器的准备和废品处理

1. 无菌仪器的准备

准备试验中常用的无菌吸管、平皿、试管、锥形瓶等时，应先将所需仪器洗涤干净，干燥后用棉花、纱布、纸等包起来，至烘箱中经160℃、2小时灭菌后方可使用，不怕水汽的仪器可用高压灭菌法灭菌。各种仪器的包扎方法如下。

（1）吸管　口吸的一端塞入少许棉花，用纸条从吸管尖端开始紧紧的将吸管全部裹缠起来，使不与外界接触。如一次大批使用吸管时，可用特制的金属桶，将大批吸管的尖端向下轻轻插入桶中，盖紧后灭菌。如一次未用完，必须重新灭菌后再用。

（2）平皿　用纸将整个平皿包起，或几个一包，亦可装入有盖的金属桶内灭菌。

（3）试管　用棉花做成棉塞，将管口塞紧后灭菌。棉塞必须松紧适宜，管内的棉塞较露出管外的棉塞长，管外的棉塞应较粗、较紧，能够将管口沿盖住。

（4）锥形瓶　用外包纱布的棉塞紧塞瓶口，再包以纱布或纸，用细麻绳扎住。

2. 废品处理

试验中被微生物污染的仪器及各种培养物，必须经过适当的处理后方可洗涤，以免微生物散布或造成感染。

（1）吸取菌液的吸管　用后轻轻插入有5%酚溶液或1%新洁尔灭溶液的玻璃桶内，浸泡1天，取出后用流水冲洗，冲洗时应不断有流水从管中通过。

（2）盛过微生物的平皿、试管、锥形瓶等一般先经高压灭菌后再用去污剂洗涤。

（3）标本片　不需保留的标本片，浸入5%的酚或1‰的新洁尔灭中1天或煮沸半小时后再洗刷，玻璃器皿洗涤后有时尚需再用清洁液浸泡。

Ⅱ. 教学用菌种

1. 细菌和放线菌

Bacillus pumilus [（CMCC（B）63202）].	短小芽孢杆菌 [CMCC（B）63202]
Bacillus subtilis	枯草杆菌
Bacillus tetani	破伤风杆菌
Bacterium typhi	伤寒杆菌
Clostridium sporogenes [CMCC（B）64941]	生孢梭菌 [CMCC（B）64941]
Clostridium tetani	破伤风梭菌
Clostridium sp	梭菌

续表

Enterobacter aerogenes	产气杆菌
Escherichia coli	大肠埃希菌
*E. coli*K12W1485	大肠埃希菌 K12W1485
*E. coli*C600	大肠埃希菌 C600
*E. coli*K12	大肠埃希菌 K12
Proteus vulgaris	普通变形杆菌
Pseudomonas aeruginosa	铜绿假单胞菌
Salmonella typhi.	伤寒沙门菌
Salmonella species	沙门氏菌
Salmonella typhimurium TA97	鼠伤寒沙门菌 TA97
Salmonella typhimurium TA98	鼠伤寒沙门菌 TA98
Salmonella typhimurium TA100	鼠伤寒沙门菌 TA100
Salmonella typhimurium TA102	鼠伤寒沙门菌 TA102
Staphylococcus aureus	金黄色葡萄球菌
Streptococcus pneumoniae	肺炎链球菌
Vibrio cholerae	霍乱弧菌
Streptomyces sp.	链霉菌

2. Fungi

Aspergillus niger	黑曲霉
Candida albicans［CMCC（F）98001］	白色念球菌［CMCC（F）98001］
Mucor	毛霉
Penicillium	青霉
Rhizopus	根霉

Ⅲ. 培养基

1. 普通肉汤琼脂培养基

牛肉膏	5g
蛋白胨	10 g
NaCl	5 g
琼脂	15 ~20 g
水	1000 ml
pH	7.41

$1kg/cm^2$高压蒸汽灭菌 20 分钟

2. 放线菌、真菌常用培养基

（1）淀粉琼脂培养基（高氏 1 号培养基）

可溶性淀粉	20 g
KNO_3	1.0 g

NaCl	0.5 g
K_2HPO_4	0.5 g
$MgSO_4$	0.5 g
$FeSO_4$	0.01 g
琼脂	15～20 g
水	1000 ml
pH	7.2～7.4

1 kg/cm^2高压蒸汽灭菌20分钟

（2）改良沙氏培养基

蛋白胨	10 g
葡萄糖	40 g
琼脂	15～20 g
水	1000 ml
pH	自然

0.56 kg/cm^2高压蒸汽灭菌15～20分钟

（3）麦芽汁琼脂培养基　大麦或小麦用水洗净，浸泡6～12小时，置15℃阴暗处发芽，上面盖一层纱布，每天用水淋3次，麦根生长至麦粒的两倍时，即可停止发芽，烘干备用。将麦芽粉碎，加4倍水，于65℃水浴中糖化3～4小时，4～6层纱布过滤。如果滤液浑浊，可用鸡蛋白澄清，即将一个鸡蛋白加水20ml，调匀至产生泡沫为止，然后倒在糖化液中搅拌煮沸后再过滤。将滤液稀释至5～6波美度，pH约为6.4，加入2%琼脂即可。

（4）葡萄糖－醋酸盐培养基

葡萄糖	1 g
酵母膏	2.5 g
醋酸钠	8.2 g
琼脂	15～20 g
水	1000 ml
pH	4.8

0.70 kg/cm^2（10磅/英寸^2，115.2℃）灭菌20分钟

3. 病毒组织培养用营养液等的配制

（1）营养液

牛血清	2.0 ml
0.5%乳蛋白水解物	97 ml
抗生素	1.0 ml

（2）维持液

0.5%乳蛋白水解物	98 ml
醋酸钠（240 mg/ml）	1.0 ml
抗生素	1.0 ml

每管加入1ml维持液，可维持6～7天。抗生素系青霉素与链霉素的混合液，每毫升

含青霉素 10000 U，链霉素 10000 U。

（3）Hanks 液

制备本溶液需用双蒸水，第二次蒸馏必须在玻璃蒸馏器内进行，所用化学药品应为分析纯试剂（A. R.），配制步骤如下：

①原液甲

NaCl	160g
KCl	8 g
$MgSO_4 \cdot 7H_2O$	2 g
$MgCl_2 \cdot 6H_2O$	2 g
$CaCl_2$	2. 8 g
双蒸水定容至	1000 ml

加 2ml 三氯甲烷作为防腐剂保存于 4℃备用

②原液乙

$Na_2HPO_4 \cdot 12H_2O$	3. 04 g
KH_2PO_4	1. 2 g
葡萄糖	20 g
0. 4% 酚红液	100 ml
双蒸水定容至	1000 ml

加 2ml 三氯甲烷作为防腐剂保存于 4℃备用

③使用时，按下列比例配成 Hanks 液　原液甲 1 份，原液乙 1 份，双蒸水 18 份，混合后于 0. 56 kg/cm^2 灭菌 10 分钟，溶液保存于 4℃，可使用 1 个月。用前于 20ml 上述液体中加入 1. 4 % $NaHCO_3$ 液 0. 5 ml（1. 4% $NaHCO_3$ 液是等渗液，制备后也以 0. 56 kg/cm^2 灭菌 10 分钟，保存于 -20℃），或把 $NaHCO_3$ 配成 3. 5% 的浓溶液，每 100 ml Hanks 液加灭菌的 3. 5% $NaHCO_3$ 1. 0 ml。

4. 各种生化反应用培养基

（1）糖发酵培养基

a. 制备"蛋白胨 - 水"培养基（蛋白胨 1%，氯化钠 0. 5%，pH7. 6）备用。

b. 配制各种单糖（葡萄糖、乳糖、麦芽糖、甘露醇、蔗糖共 5 种）的 20% 水溶液，0. 56 ~ 0. 7kg/cm^2 高压灭菌 20 分钟。

c. 取上述蛋白胨 - 水培养基 100ml，加 1. 6% 溴甲酚紫 0. 1ml，混匀，分装于小试管中，内倒置一杜氏小管，0. 56 ~ 0. 7kg/cm^2 高压灭菌，以无菌操作加入相应的灭菌单糖溶液于每管中，使最终浓度为 0. 5% ~ 1. 0%。

（2）"磷酸盐 - 葡萄糖 - 蛋白胨 - 水"培养基（甲基红 - VP 试验培养基）

蛋白胨	0. 5g
磷酸氢二钾	0. 5g
葡萄糖	0. 5g
蒸馏水	100ml
pH	7. 2 ~ 7. 6

0. 56 ~ 0. 7kg/cm^2 高压灭菌 20 分钟

（3）枸橼酸盐琼脂培养基

枸橼酸钠（无水）	0.2 g
氯化钠	0.5 g
硫酸镁	0.02 g
磷酸氢二钾	0.1 g
磷酸二氢铵	0.1 g
琼脂	2 g
蒸馏水	100ml

以上成分加热融化，调 pH 至 6.8～7.0，加入 1% 溴麝香草酚蓝 1 ml，混匀，分装试管，经 0.7kg/cm^{2}高压灭菌 20～30 分钟后制成斜面即可。

（4）蛋白胨－水培养基（吲哚试验用）

蛋白胨	1 g
氯化钠	0.5 g
蒸馏水	100ml

pH7.2～7.4

1 kg/cm^{2}高压灭菌 20 分钟

（5）醋酸铅培养基

1.5%～2.0% 普通肉汤琼脂培养基	100 ml
硫代硫酸钠	0.25 g
10 % 醋酸铅溶液	1 ml

将琼脂培养基加热融化，冷却至约 60℃，加入硫代硫酸钠，混合，高压灭菌。冷却至约 50℃，无菌操作加入醋酸铅溶液，混匀，分装试管，每管 3～5ml，直立待凝即可。醋酸铅溶液预先经 0.56 kg/cm^{2}高压灭菌 15 分钟。

（6）明胶培养基　肉汤培养基 100ml 加入明胶 12～18g，加热融化，分装试管，每管 2～3ml，0.56 kg/cm^{2}高压灭菌 30 分钟，最终 pH 为 7.2～7.4，直立待凝。

（7）淀粉琼脂培养基

琼脂	12.0g
可溶性淀粉	10.0 g
牛肉膏	13.0g
pH	7.5

1 kg/cm^{2}高压灭菌 20 分钟

（8）尿素琼脂培养基

尿素	20.0 g
琼脂	15 g
氯化钠	5.0g
磷酸二氢钾	2.0g
蛋白胨	1.0g
葡萄糖	1.0g
酚红	0.012g

5. 抗生素微生物检查用培养基

（1）培养基Ⅰ

胨	5g
牛肉浸出粉	3 g
磷酸氢二钾	3 g
琼脂	15 ~ 20g
水	1000 ml

除琼脂外，混合上述其余成分，调节 pH 使比最终的 pH 略高 0.2 ~ 0.4，加入琼脂，加热溶化后过滤，调节 pH 使灭菌后为 7.8 ~ 8.0 或 6.5 ~ 6.6，115℃灭菌 30 分钟。

（2）培养基Ⅱ

胨	6 g
牛肉浸出粉	1.5 g
酵母浸出粉	6g
葡萄糖	1 g
琼脂	15 ~ 20g
水	1000 ml

除琼脂和葡萄糖外，混合上述其余成分，调节 pH 使比最终的 pH 略高 0.2 ~ 0.4，加入琼脂，加热溶化后过滤，加葡萄糖溶解后摇匀，调节 pH 使灭菌后为 7.8 ~ 8.0 或 6.5 ~ 6.6，115℃灭菌 30 分钟。

6. 无菌检查用培养基

（1）需气菌、厌气菌培养基（硫乙醇酸盐流体培养基）

酪胨	15.0 g
葡萄糖	5 g
L－胱氨酸	0.5 g
硫乙醇酸钠（或硫乙醇酸 0.3 m l）	0.5 g
酵母浸出粉	5.0 g
氯化钠	2.5 g
0.1% 刃天青（新配制）	1.0 ml
琼脂粉	0.5 ~ 0.75g
水	1000 ml

除葡萄糖和刃天青溶液外，取上述其余成分加入水中，微温溶解，调节 pH 至弱碱性，煮沸后过滤，加葡萄糖和刃天青溶液，摇匀，调节 pH 使灭菌后为 7.1 ± 0.2，分装，灭菌。

（2）真菌培养基

胨	5.0 g
酵母浸出粉	2.0 g
葡萄糖	20 g
磷酸氢二钾	1g
硫酸镁	0.5g

水	1000 ml

除葡萄糖外，取上述其余成分加入水中，微温溶解，调节 pH 约为6.8，煮沸，加葡萄糖溶解后摇匀，调节 pH 使灭菌后为6.4 ± 0.2，分装，灭菌。

7. 微生物限度检查用培养基

（1）玫瑰红钠琼脂培养基

胨	5.0 g
葡萄糖	10.0 g
磷酸二氢钾	1 g
硫酸镁	0.5g
玫瑰红钠	0.0133g
琼脂	15 ~ 20 g
水	1000 ml

除葡萄糖、玫瑰红钠外，取上述其余成分混合，加热溶化后，过滤，加葡萄糖、玫瑰红钠后分装，灭菌。

（2）酵母浸出粉胨葡萄糖琼脂培养基

胨	10.0 g
酵母浸出粉	5g
葡萄糖	20.0 g
琼脂	15 ~ 20 g
水	1000 ml

除葡萄糖外，取上述其余成分混合，加热溶化后，过滤，加葡萄糖后分装，灭菌。

（3）胆盐乳糖培养基

胨	20.0 g
乳糖	5g
氯化钠	5 g
磷酸氢二钾	4.0g
磷酸二氢钾	1.3g
去氧胆酸钠	0.5g
水	1000 ml

除乳糖、去氧胆酸钠外，取上述其余成分混合，加热溶化后，过滤，加乳糖、去氧胆酸钠后分装，灭菌。

（4）曙红亚甲基蓝琼脂培养基

营养琼脂培养基	100ml
20%乳糖溶液	5ml
曙红钠指示剂	2ml
亚甲基蓝指示剂	1.3 ~ 1.6ml

取营养琼脂培养基，加热溶化后，冷却至60℃，按无菌操作加入灭菌的其他3种溶液，摇匀，倾注平板。

Ⅳ. 染色液的配制

1. 常用染料的溶解度

染料	溶于100ml溶剂的染料量（g）	
	水	95%乙醇
结晶紫	1.68	13.87
伊红	44.2	2.18
碱性复红	1.03	3.20
孔雀绿	7.60	7.52
美蓝	3.55	1.48
沙黄（番红）	5.45	3.41
甲苯胺蓝	3.82	0.57

一般配法系将染料加入到玻璃研磨内，加入95%乙醇后放置孵箱内数日（每日振摇数次），然后用滤纸过滤备用。

2. 革兰染液

（1）结晶紫染液　结晶紫14g，溶于95%乙醇100ml中使成饱和液备用，取该饱和液200ml与1%草酸胺水溶液80ml混合即成。

（2）路哥碘液　碘化钾2g溶于100ml蒸馏水中，再加碘1g，至全溶后徐徐加蒸馏水至300ml即成。

（3）稀释复红溶液　用蒸馏水将石炭酸复红染液（如下）稀释10倍，即为稀释复红染液。

3. 鞭毛染色液

（1）鞭毛染色液甲　将明矾饱和液20ml、20%鞣酸10ml、95%乙醇15ml、蒸馏水10ml和复红酒精饱和液3ml混合即可。

（2）鞭毛染色液乙　1.0 g硼砂、1.0 g美蓝溶于200ml蒸馏水中。

4. 抗酸染色液

（1）石炭酸复红染液　碱性复红4g溶于100ml 95%的乙醇中制成饱和液备用，取该饱和液10ml与5%石炭酸溶液90ml混匀即成。

（2）3%盐酸酒精　取浓盐酸3ml缓缓滴入97ml 95%的乙醇中。

（3）吕氏美蓝染液　取美蓝2g溶于100ml 95%乙醇中使成饱和溶液备用，取此饱和液30ml与0.01%的KOH水溶液100ml混匀即成。

5. 方他那染液的配制

（1）方他那甲液　冰醋酸1ml和40%甲酸2ml溶于100ml蒸馏水。

（2）方他那乙液　鞣酸5g加热溶于100ml蒸馏水中。

（3）方他那丙液　硝酸银4g，溶于200ml蒸馏水中，取50ml硝酸银溶液，滴加浓氨水，使先有沉淀发生复而溶解，然后再加入硝酸银溶液少许，至振荡后仍显轻度浑浊为止。澄清的溶液不能使用。

V. 试剂及缓冲液配制

1. 常用指示剂的配制

称取指示剂0.1g，置于研钵中磨成粉末，按下表所示滴加0.01mol/L NaOH，加蒸馏水至规定量（250 ml）即可。

指示剂	色调变更 酸→碱	稀释0.1g指示剂所需0.01mol/L NaOH量（ml）	总体积（ml）	浓度（%）	10ml培养基所需量（ml）
溴甲酚蓝	黄→紫	1.85	250	0.04	0.5
溴酚蓝	黄→蓝	1.49	250	0.04	0.5
溴麝香草酚蓝	黄→蓝	1.60	250	0.04	0.5
甲基红	红→黄	-	500	0.02	0.5
酚红	黄→红	2.82	500	0.02	0.5
麝香草酚蓝	黄→蓝	2.15	250	0.04	0.5

2. 各种生化反应用试剂的配制

（1）甲基红试剂　取甲基红0.1g，溶于300ml 95%乙醇中，蒸馏水定容至500ml。

（2）寇氏试剂　5.0g对二甲基氨基苯甲醛加至75 ml戊醇中，置50~60℃水浴中，搅拌使之完全溶解，冷却，将25 ml浓盐酸一滴滴徐徐加入，边加边摇，配制后装棕色瓶并放暗处保存。

APPENDIX

Ⅰ. Sterilization of Glass Apparatus and Treatment of Waste

1. Preparation and Sterilization of Glass Apparatus

Sucker, petri dish, tubes and flask should be washed before they are desiccated. The desiccated vessels should be packed with cotton, gauze or paper, placed in oven and heated 2 hours under 160℃. The apparatus which can tolerate vapor can be autoclaved. The methods of package of apparatus are these:

(1) Sucker stuff a little cotton in the end of sucker, wrap the sucker with paper strip, heat 2h under 160℃.

(2) Petri dish package the petri with paper and heat 2h under 160℃.

(3) Tube stuff the jaws of tube with silica gel tuck and heat 2h under 160℃.

(4) Flask stuff the jaws of tube with silica gel tuck and heat 2h under 160℃.

2. Treatment of Waste

the culture and apparatus, which contaminated microbial should be treated before they can be cleaned.

(1) Sucker treat with 5% hydroxybenzene a day, wash it with water.

(2) Petri, tube and flask autoclave.

(3) Specimen treat with 5% hydroxybenzene a day.

Ⅱ. Microorganisms

1. Bacteria and Actinomycetes

Bacillus pumilus [CMCC (B) 63202]

Bacillus subtilis

Bacillus tetani

Bacterium typhi

Clostridium sporogenes [CMCC (B) 64941]

Clostridium tetani

Clostridium sp

Enterobacter aerogenes

Escherichia coli

E. coli K12W1485

E. coli C600

E. coli K12

Proteus vulgaris

Pseudomonas aeruginosa

Salmonella typhi

Salmonella species

Salmonella typhimurium TA97

Salmonella typhimurium TA98

Salmonella typhimurium TA100

Salmonella typhimurium TA102

Staphylococcus aureus

Streptococcus pneumoniae

Vibrio cholerae

Streptomyces sp.

2. Fungi

Aspergillus niger

Candida albicans [CMCC (F) 98001]

Mucor

Penicillium

Rhizopus

Ⅲ. Culture Media

1. Nutrient Broth Agar

Beef extract	5 g
Peptone	10 g
NaCl	5g
Agar	15 ~ 20 g
Water	1000 ml
pH	7. 4

Autoclave for 20min at 1 kg/cm^2.

2. Media for Actitomycetes and Fungi

(1) Gauss′ agar

Soluble starch	20 g
KNO_3	1.0 g
NaCl	0.5 g
K_2HPO_4	0.5 g
$MgSO_4$	0.5 g
$FeSO_4$	0.01 g
Agar	15 ~ 20 g
H_2O	1000 ml
pH	7.2 ~ 7.4

Autoclave for 20min at 1 kg/cm^2

(2) Improved sabroud medium

Peptone	10 g
Glucose	40 g
Agar	15 ~ 20g
water	1000 ml
pH	natural

Autoclave for 15 ~ 20min at 0.56 kg/cm^2.

(3) Wort agar

Agar	15g
Malt extract	15g
Maltose	12.75g
Dextrin	2.75g
Glycerol	2.35g
K_2HPO_4	1g
NH_4Cl	1g
Pancreatic digest of gelatin	0.78g
pH	4.8

Autoclave for 15 ~ 20min at 0.56 kg/cm^2.

Preparation of Medium: Add components to distilled/deionized water and bring volume to 1.0 L. Mix thoroughly. Gently heat and bring to boiling. Boil for 1 min with mixing. Distribute into tubes or flasks. Autoclave for 15 min at 15 psi pressure or 121℃. Do not overheat as this will result in hydrolysis of the agar. An additional 5.0g of agar can be used to make a firmer agar. Pour into sterile Petri dishes or leave in tubes Reaction: The low pH of the agar selectively inhibits bacterial growth and permits the growth of yeast.

(4) glucose - acetate agar

Glucose	1 g
Yeast extract.	2.5g
Natrium acetate	8.2 g
Agar	15 ~ 20 g
Water	1000 ml
pH	4.8

Autoclave for 20min at 0.7kg/cm^2.

3. virus Culture Broth

(1) nutrient broth

Beef serum	2.0 ml
0.5% milk protein	97 ml
Antibiotin solution	1.0 ml

(2) maintain solution

0.5% digest of casein	98 ml
Natrium acetate (240 mg/ml)	1.0 ml
Antibiotin solution	1.0 ml

1ml maintain solution per tube. Antibiotic solution contains penicilium 10000 U and streptomycin 10000 U.

(3) Hank's solution

(a) Solution A

NaCl	160g
KCl	8 g
$MgSO_4 \cdot 7H_2O$	2 g
$MgCl_2 \cdot 6H_2O$	2 g
$CaCl_2$	2.8 g
ddH_2O	1000 ml

Add chloroform 2ml to the solution. preserved at 4℃.

(b) Solution B

$Na_2HPO_4 \cdot 12H_2O$	3.04 g
KH_2PO_4	1.2 g
Glucose	20 g
0.4% bromcresol blue	100 ml
ddH_2O	1000 ml

Add chloroform 2ml to the solution. preserved at 4℃.

(c) Preparation Add solution A 10ml and solution B 10ml to ddH_2O 18ml, mixed and autoclave for 20min at 0.56kg/cm^2. The solution can be preserved in 4℃.

Before use, add 0. 5ml 1. 4 % $NaHCO_3$.

4. media for Biochemistry Test

(1) Carbohydrate fermentation medium.

(a) Peptone – water medium (g/100ml): (peptone 1g, nacl 0. 5g, pH7. 6).

(b) Sugars solutions: 2g glucose (or lactose, maltose, mannitol, sucrose) in 10ml water. autoclave for 20min at 0. 56kg/cm^2.

(c) In the tubes of sterile peptone – water medium (about 5ml) with inverted durham tubes, Add 10μl 1. 6% phenol red to the medium. autoclave for 20min at 0. 56kg/cm^2. Add 200μl 20% sugar to every medium.

(d) MRVP broth

Peptone	0. 5g
KH_2PO_4	0.5g
Glucose	0. 5g
Distilled water	100ml
pH	7. 2 ~ 7. 6

Autoclave for 20min at 0. 56 ~ 0. 7kg/cm^2

(2) Simmons' citrate agar

Citrate	0. 2 g
NaCl	0. 5 g
$MgSO_4$	0. 02 g
K_2HPO_4	0. 1 g
$NH_4H_2PO_4$	0. 1 g
Agar	2 g
Distilled water	100ml

Melt the components, adjust the pH to 6. 8 ~ 7. 0. Add bromthymol blue 1 ml to the solution, distribute to tubes. autoclave for 20min at 0. 7kg/cm^2.

(3) Peptone – water medium (for indole test)

Peptone	1 g
NaCl	0. 5 g
Distilled water	100ml
pH	7. 2 ~ 7. 4

Autoclave for 20min at 1. 0kg/cm^2.

(4) Lead acetate agar

1. 5% ~ 2. 0% broth agar	100 ml
$Na_2S_2O_3$	0. 25 g
10 % lead acetate solution	1 ml

Melt the broth agar, cooling to 60℃. Add $Na_2S_2O_3$ to solution, mix solution, autoclave for 20min at 1.0kg/cm^2. Cooling to 50℃, add 10 % lead acetate solution to broth agar, distribute to tubes (3ml per tube), erect the tubes, let the broth agar freeze to the solid state.

(5) glutin medium add glutin 12 ~ 18g to 100ml broth, melt the broth, distribute to tubes (3ml per tube), autoclave for 20min at 1.0kg/cm^2, erect the tubes, let the broth agar freeze to the solid state.

(6) starch agar (g/L)

Agar	12.0g
Soluble starch	10.0g
Beef extract	3.0g
pH	7.5

Autoclave for 10min at 1.0kg/cm^2.

(7) Urea agar (g/L)

Urea	20.0g
Agar	15.0g
NaCl	5.0g
KH_2PO_4	2.0g
Peptone	1.0g
Glucose	1.0g
Phenol red	0.012g
pH	6.8

Autoclave for 15min at 1.0kg/cm^2.

5. Media for Biological Assay of Antibiotics

(1) Medium A

Peptone	6 g
Beef extract	3 g
Dipotassium hydrogen phosphate	3 g
Agar	15 ~ 20g
Water to produce	1000 ml

Dissolve the solids in the water except agar, Add 1M sodium hydroxide so that the value of solution pH is 0.2 ~ 0.4 higher than the value of sterilized medium. Add the agar, boil and filtrate the solution, adjust the pH so that the solution will have a pH of 7.8 ~ 8.0 or 6.5 ~ 6.6 after sterilization. Sterilize the medium at 115℃ for 30 min.

(2) Medium B

Peptone	6 g
Beef extract	1.5 g
Yeast extract	6 g

Glucose monohydrate	1g
Agar	15 ~ 20 g
Water to produce	1000 ml

Dissolve the solids in the water except glucose and agar, Add 1M sodium hydroxide so that the value of solution pH is 0. 2 ~ 0. 4 higher than the value of sterilized medium. Add the agar, boil and filtrate the solution, add the glucose, shake up and adjust the pH so that the solution will have a pH of 7. 8 ~ 8. 0 or 6. 5 ~ 6. 6 after sterilization. Sterilize the medium at 115℃ for 30 min.

6. media for Sterility Test

(1) Aerobes and anaerobes medium (Fluid thioglycollate medium)

Pancreatic digest of casein	15. 0 g
Glucose monohydrate	5g
L – Cystine	0. 5 g
Sodium thioglycollate 0. 5 gor Thioglycollic acid	0. 3 ml
Yeast extract (water – soluble)	5. 0 g
Sodium chloride	2. 5 g
Resazurin sodium solution (1 in 1000)	1. 0 ml
Freshly prepared,	
Agar, granulated	0. 75 g
(moisture content not in excess of 15%)	
Water	1000 ml
pH 7. 1 ± 0. 2 after sterilization	

Dissolve the solids in the water except glucose and resazurin sodium solution, warm slightly to effect solution. Add 1M sodium hydroxide so that the solution will be weakly basic. Boil and filtrate the solution, add the glucose and resazurin sodium solution, shake up and adjust the pH so that the solution will have a pH of 7. 1 ±0. 2 after sterilization. Distribute and sterilize the medium.

(2) Fungi medium

Peptone	5. 0 g
Yeast extract	2. 0 g
Glucose	20 g
Dipotassium hydrogen phosphate	1g
$MgSO_4$	0. 5g
Water	1000 ml
pH 6. 4 ± 0. 2 after sterilization	

Dissolve the solids in the water except glucose, warm slightly to effect solution. Add 1M sodium hydroxide so that the solution will have a pH of 6. 8. Boil the solution and add the glucose, adjust the pH so that the solution will have a pH of 6.4 ±0.2 after sterilization.

7. media for Microbial Contamination Test

(1) Rose Bengal agar

Peptones (meat and casein)	5.0 g
Glucose	10.0 g
Dipotassium hydrogen orthophosphate	1 g
Magnesium chloride	0.5g
Rose Bengal Sodium	0.0133g
Agar	15 ~ 20 g
Water	1000 ml

Dissolve the solids in the water except glucose and rose bengal sodium, warm slightly to effect solution. Filtrate the solution, add glucose and rose bengal sodium, sterilize the medium.

(2) Yeast extract peptone dextrose medium (YPD)

Peptones (meat and casein)	10.0 g
Yeast extract	5g
Glucose	20.0 g
Agar	15 ~ 20 g
Water	1000 ml

Dissolve the solids in the water except glucose, warm slightly to effect solution. Filtrate the solution, add glucose and sterilize the medium.

(3) Bile salt lactose medium (BL)

Peptones (meat and casein)	20.0 g
Lactose	5g
Sodium chloride	5 g
Dipotassium hydrogen phosphate	4.0g
Potassium dihydrogen phosphate	1.3g
Bile salt Sodium (or deoxycholate 0.5g)	2g
Water	1000 ml

Dissolve the solids in the water except lactose and sodium deoxycholate, warm slightly to effect solution. Add 1M sodium hydroxide so that the solution will have a pH of 7.4 ±0.2 after sterilization. Boil and filtrate the solution, add lactose and sodium deoxycholate, sterilize the medium.

(4) EMB

Nutrient Broth	100ml
20% Lactose solution	5ml

Eosin	2ml
Methylene blue solution	1. 3 ~ 1. 6ml

Melt the nutrient broth agar and cool to 60℃, add the other solutions by aseptic technology, shake up and sterilize. Pour the petri disks.

Ⅳ. Staining Solutions

1. Solubility of Dye

Dyes	Solubility (g/100ml solvent)	
	Water	95% Ethyl alcohol
Crystal violet	1. 68	13. 87
Eosin	44. 2	2. 18
Fuchsin basic	1. 03	3. 20
Malachite green	7. 60	7. 52
Methylene blue	3. 55	1. 48
Safranin	5. 45	3. 41
Toluiden blue	3. 82	0. 57

Preparation: grounded in the mortar, the dye was dissolved with 95% ethanol under 37℃ (shake every day). Before use, the solution must be filtrated with filter paper.

2. Gram's Stain Solutions

(1) Crystal violet solution A: 2g crystal violet in 20ml 95% ethanol. solution B: 0. 8g ammonium oxalate in 80ml distilled water. Mix solution A and solution B and let stand for 24hours. filter through paper.

(2) Iodine solution 1g iodine and 2g potassium iodide in 300ml water.

(3) Safranin 2. 5g safranin in 100ml of 95% ethyl alcohol. Add 1000ml of distilled water.

3. Flagella Stain Solutions

(1) Solution A saturated $KAl(SO_4)_2$ solution 20ml, 20% tannic acid solution 10 ml, 95% ethanol 15 ml, distilled water 10 ml and saturated safranin solution 3 ml. Mix thoroughly.

(2) Solution B boron 1. 0 g and methylene blue 1. 0 g in distilled water 200ml.

4. Acid – Fast Stain Solution

(1) Carbol fuchsin stain basic fuchsin 0. 4g in 95% ethanol 10ml. Then mix with 5% phenol 90ml.

(2) 3% acid alcohol concentrated HCl 3ml in 97ml 95% ethanol.

(3) Methylene blue solution 2g methylene blue in 100ml 95% ethanol. Add the solution 30ml to 0. 01% KOH 100ml.

5. Fontana Solution

(1) Fontana solution A acetate 1ml and 40% formic acid 2ml in 100ml distilled water.

(2) Fontana solution B tannic acid 5 g in 100 ml distilled water.

(3) Fontana solution C $AgNO_3$ 4g in 200ml distilled water. add concentrated NH_3 to the 50 ml $AgNO_3$ solution, the precipitation will be produced, then add a little $AgNO_3$ solution, shake the solution to leave little precipitation in it.

V. Solutions and Buffers

1. Indicators

0.1g grinded indicator dissolved in 0.01mol/L NaOH, add distilled water to 250 ml.

Indicator	Color change Acid →basic	Volume of NaOH to dissolve indicator (ml)	Total volume	Concentration (%)	Volume in 10ml medium (ml)
bromcresol blue	yellow→violet	1.85	250	0.04	0.5
bromphenol blue	yellow→blue	1.49	250	0.04	0.5
bromthymol blue	yellow→blue	1.60	250	0.04	0.5
methyl red	red→yellow	-	500	0.02	0.5
phenol	yellow→red	2.82	500	0.02	0.5
dresolphthalein	yellow→blue	2.15	250	0.04	0.5

2. Sulutions for Biochemistry Test

(1) Methyl red indicator methyl red 0.1g is dissolved in 300ml 95% ethanol. Add distilled water to 500ml.

(2) Kovac's reagent 5.0g dimethyl - alpha - naphthylamine is dissolved in 75ml amyl alcohol. Heat at 55℃ in water bath until crystal is dissolved. Cool and add 25ml concentrated hcl. Preserve it in umber bottle.

Words and Vocabulary

α – naphthol solution	α – 萘酚
acetate agar	醋酸琼脂
acetate buffer	醋酸缓冲液
acid and gas production	产酸产气
acid – fast bacteria	抗酸细菌
acid – fast stain	抗酸染色法
acridine orange stain	吖叮黄
actinomycete	放线菌
aerial filament	气生菌丝
aerobes	需氧微生物
aerosol	气雾剂
agar	琼脂
agarose gel	琼脂糖
agglutination	凝聚反应
air dry	空气干燥
ames test	艾姆斯实验
amplification	扩增
amylase	淀粉酶
anaerobes	厌氧微生物
anaerobic bacteria	厌氧细菌
anaerobic culture techniques	厌氧培养技术
antibiotic	抗生素
antimicrobial activity	抗菌活性
antimicrobial drug spectrum	抗菌谱
antibiotic medium	抗生素培养基
antibiotic resistance	抗生素抗性
antibiotic – resistant mutants	抗生素抗性突变
antiseptic	防腐
annulus	环状物
asepsis	无菌
aseptic	无菌的

aseptic technique	无菌技术
Aspergillus	曲霉菌
autoclave	高压灭菌
auxanogram	生长谱
auxanography	生长谱法
azithromycin	阿奇霉素
Bacillus	杆菌
Bacillus pumilus	短小芽孢杆菌
Bacillus tetani	破伤风杆菌
back mutation	回复突变
bacteria	细菌
bacterial classification	细菌分类
morphology	形态
bacterial endospore stain	细菌芽孢染色
bacterial identification	细菌鉴定
bacterial reproduction	细菌繁殖
bacterial slant	细菌斜面形态
bacterial stab	细菌穿刺
bacterial transformation	细菌转化
bacteriophage	噬菌体
bacteriophage enrichment	噬菌体富集
barbital buffer	巴比妥缓冲液
beta lactam	β－内酰胺环
barritt's reagent	Barritt 试剂
basal synthetic medium	基础合成培养基
basophils	嗜碱性细胞
bauer－kirby disk method	B－K 法
beef extract	牛肉膏
binary fission	二分裂
biomass	生物量
BL	胆盐乳糖
blood agar	血琼脂
blood type	血型
brewer jar	广口瓶
bright－field microscope	亮视野显微镜
bromcresol purple broth	溴甲酚紫肉汤
brown' s motion	布朗运动
broth medium	肉汤培养基
calcium chloride	氯化钙

Candida albicans, CMCC（F）98001 白色念珠菌
capsule 荚膜，胶囊剂
capsule stain 荚膜染色
carbohydrate fermentation 糖发酵实验
carbol fuchsin 石炭酸品红
carbon source 碳源
carcinogen 致癌物质
casein hydrolysate 水解酪蛋白
catabolite repression 代谢物抑制
catalase 触酶，过氧化氢酶
carcinogen 致癌因子，致癌剂
carcinogenic 致癌的
carcinogenicity 致癌性
cedar wood oil 香柏油
cell wall 细胞壁
cellulose decomposition 纤维素降解
centrifugation 离心
centrifuge 离心，离心机
cesium chloride 氯化铯
chloramphenicol（cm） 氯霉素
chromogen 发色基团
chromosome 染色体
chlamydia 支原体
citrate agar 枸橼酸盐琼脂
citrate utilization 枸橼酸盐利用
chook embryos 鸡胚
Clostridium 梭菌
Clostridium sporogenes 生孢梭菌
colony 菌落
colony forming unit（CFU） 菌落形成单位
coliform bacteria 肠道菌
competent bacterium 感受态细菌
competent cells 感受态细胞
complex medium 合成培养基
concavity slide 凹玻片
conjugation 接合
contamination 污染
continuous streak 连续划线
correlation 相关性

counterstain	复染
counting chamber	血球计数板
counting chamber	计数室
cover slip	盖玻片
crystal violet stain	结晶紫
crossover	交换
culture	培养物
cultivation of bacteria	细菌培养
culture	培养
culture morphology	培养状态
culture preservation	培养物保存
cysteine desulfurase	半胱氨酸脱硫酶
cytochrome oxidase	细胞色素氧化酶
dark – field microscope	暗视野显微镜
daunomycin	正定霉素
deaminase	脱氨酶
decarboxylase	脱羧酶
decolorizing	脱色
decolorizing agent	脱色剂
decline or death phase	衰退期
stationary phase	稳定期
defined medium	限制性培养基
degrade	降解
detergent	去污剂
diethyl sulfate (des)	硫酸二乙酯
disk diffusion	平板扩散
donor	供体
dropper	滴管
drug – resistant	抗药性
denitrification	脱氮
desiccation	干燥
deoxycholate agar	双氧胆酸盐琼脂
desoxycholate – citrate agar (dca)	双氧胆酸盐 – 枸橼酸盐琼脂
dextran	葡聚糖
differential media	鉴别培养基
direct counts	直接计数
dimethylsulfoxide	二甲亚砜
disinfectant	消毒剂
disinfectant effectiveness	消毒剂效果

disinfectants	消毒剂
donor DNA	供体 DNA
dry heat sterilization	干热灭菌
ear drops	滴耳剂
electrophoresis	电泳
EMB agar	EMB 琼脂
endospore stain	芽孢染色
emulsions	乳剂
enteric gram – negative rods	革兰阴性肠道菌
Enterobacter aerogenes	产气肠杆菌
Enterobacter species	产气杆菌属
environmental microbiology	环境微生物
exonuclease	内切酶
enzyme assays	酶分析
enzyme induction	酶诱导
enzyme – linked immunosorbent assay	酶联免疫分析
eosin methylene blue agar	依红美蓝琼脂
eosinophils	曙红
error – prone	倾向差错的
Escherichia coli	大肠埃希菌
ethidium bromide staining	EB 染色
experimental design	实验设计
eye drops	一般滴眼剂
eye ointments	一般眼膏剂
eukaryote	真核细胞
F – pili	性菌毛
F – plasmid	F 质粒
facultative anaerobes	专性厌氧菌
Fermentation	发酵
films	膜剂
fixation	固定
flagella	鞭毛
flagella stain	鞭毛染色
fluorescein – conjugated antibody	荧光结合抗体
fluorescent antibody staining (fab)	荧光抗体染色
freeze – drying	冷冻干燥
fuchsin	复红
fungus	真菌
fungal classification	真菌分类

galactose	半乳糖
gel	凝胶剂
gelatin hydrolysis	明胶水解
gelatinase	明胶酶
generalized transduction	普通转导
genome	基因组
germicides	杀菌剂
glass slides	载玻片
glycolysis	糖酵解
gonococcus	淋病双球菌
gram stain	革兰染色
gram – negative	革兰阴性
gram – positive	革兰阳性
granules	颗粒剂
growth curve	生长曲线
hand washing	手洗
hanging drop method	悬滴法
heat killing	加热杀死
heavy metals	重金属
histidine	组氨酸
Hfr strain	高频重组菌株
homogenate	匀浆
hydrogen sulfide production	硫化氢产生实验
immune response	免疫应答
immunofluorescence	免疫荧光
immunoglobulins	免疫球蛋白
immunology	免疫学
incubation time	接种时间
incubator	培养箱
indicator microorganisms	指示微生物
indirect counts	间接计数
indirect immunofluorescence	降解免疫荧光反应
indole production	吲哚产生实验
inoculate	接种
inhibition zone	抑菌圈
inoculating loop	接种环
inoculating needle	接种针
inoculum	接种量
insecticide	杀虫剂

integration 整合
irradiation 照射
isoelectric point 等电点
isolation of bacteriophage 噬菌体分离
kirby – bauer antibiotic susceptibility test K – B 法
Klebsiella 克雷伯菌
Klebsiella. pneumoniae 肺炎克雷伯菌
kovac's reagent 寇氏试剂
lactose 乳糖
lactose – fermenting bacteria 乳酸发酵细菌
lag phase 延迟期
lead acetate agar 醋酸铅琼脂
lens paper 擦镜纸
leukocytes 白细胞
levofloxacin 左旋氧氟沙星
liniments 搽剂
liquid paraffin 液体石蜡
liquid nitrogen 液氮
liquid medium 液体培养基
light microscope 光学显微镜
logarithmic (log) phase 对数生长期
lotions 洗剂
lyophilizing 冻干
lysozyme 溶菌酶
ligase 连接酶
lysogenic 溶源的
lysate 裂解物
lytic bacteriophage 烈性噬菌体
lytic phage replication cycle 烈性噬菌体复制循环
Macconkey agar 麦康凯培养基
malt extract 麦芽汁
mating 杂交
mean 平均
medium 培养基
medium preparation 培养基制备
membrane filter method 膜过滤技术
meningococcus 脑膜炎球菌
metabolism 代谢
methyl red test 甲基红实验

methyl red – voges – proskauer broth	MRVP 肉汤
microaerophiles	微厌氧菌
microbial biomass	生物量
microbial classification	微生物分类
microscopic counts	显微计数
microbial genetics	微生物遗传
microbial growth	微生物生长
microbial identification	微生物鉴定
microbial metabolism	微生物代谢
microbial nomenclature	微生物命名
microbial taxonomy	微生物分类
microbiological methods	微生物法
micrometer	测微尺
microscope	显微镜
minimal agar	基本培养基琼脂
MBC	最低杀菌浓度
minimal bactericidal concentration	最低杀菌浓度
minimal broth	基本培养基肉汤
MIC	最低抑菌浓度
minimal inhibitory concentration	最低抑菌浓度
mite	螨
minimal medium（MM）	最低限度培养基
mold	霉菌
mordant	媒染
morphology	形态学
most probable number method	MPN 法
Mucor	毛霉
M – H	M – H 培养基
Mueller – Hinton agar	M – H 琼脂
mutagen	突变剂
mutant	突变
mycelia	菌丝
mycobacterium	分枝杆菌
mycoplasma	支原体
nasal drops	滴鼻剂
neutrophils	中性粒细胞
nitratase	硝酸还原酶
2 – nitrofluorene	2 – 氨基芴
nitrification	硝化作用

nitrogen fixation	固氮
normal growth curve	常规生长曲线
NTG	亚硝基胍
nucleotide	核苷酸
mycostatin	制霉菌素
numerical aperture	数值口径
nutrient agar	营养琼脂
nutrient agar slant	营养琼脂斜面
nutrient broth	营养肉汤
oil immersion	油镜
ointments	软膏剂
oral liquids	口服液
paper slips	纸片
Pasteurization	巴氏消毒
pathogenic bacteria	致病菌
penicillin	青霉素
penicillin G	青霉素 G
penicillinase	青霉素酶
penicilloic acid	青霉噻唑酸
Penicillium	青霉菌
petri dish	培养皿
pharmacopoeia	药典
physiological saline	生理盐水
pills	丸剂（滴丸、糖丸等）
plasmid DNA	质粒 DNA
plasmid	质粒
Pneumococcus pneumoniae	肺炎球菌
polypeptone	聚胨
polysaccharides	多糖
polypeptides	多肽
potato dextrose agar	PDA 培养基
pour plate method	倾注平板法
powders	散剂
powders for external use	外用散剂
precipitin lines	沉淀线
primary stain	初染
procaryote	原核细胞
proteose peptone	蛋白胨
Proteus species	变形杆菌

protoplasts	原生质体
prophage	前噬菌体
pure culture	纯培养
Pseudomonas aeruginosa	铜绿假单胞菌
Pseudomonas species	假单胞菌
pure culture	纯培养
pure culture techniques	纯培养技术
puncture	穿刺
quadrant streak	分区划线
recombinant DNA technology	重组 DNA 技术
replica plating	影印平板法
Rhizopus	根霉
Rickettsiaes	立克次体
Rose Bengal agar	玫瑰红钠琼脂
ruler	尺
sabouraud agar	沙氏琼脂
Saccharomyces cerevisiae	啤酒酵母
safranin	沙黄
Salmonella typhimurium	伤寒沙门菌
Salmonella	沙门菌
Salmonella typhi	伤寒杆菌
Sarcina lutea	藤黄八叠球菌
screen	筛选
smear	涂片
solid medium	固体培养基
stain	染色
staining tray	染缸
stationary phase	稳定期
selective media	选择培养基
serological identification	血清学鉴定
semisolid medium	半固体培养基
Shigela species	志贺菌
Simmons' citrate agar	西蒙枸橼酸盐琼脂
simple staining	单染法
soya peptone	大豆蛋白胨
spectrophotometric	光比色法
specialized transduction	局限性转导
specimen	标本片
spheroplasts	球形体

spirochetes	螺旋体
spontaneous mutations	自发突变
spore	芽孢
sporebearing filament	孢子丝
spore stain	芽孢染色
spread plate method	平板划线法
staining	染色
standard error	标准差
Staphylococcus	葡萄球菌
Staphylococcus aureus	金黄色葡萄球菌
Staphylococcus. epidermidis	表皮葡萄球菌
starch	淀粉
starch agar	淀粉琼脂
starch agar slant	淀粉琼脂斜面
starch hydrolysis	淀粉水解
statistics	统计学
sterilization	灭菌
streak plate	划平板
streptomycin	链霉素
subculturing	接种
substrate filament	基内菌丝
sucker	吸管
sudan black stain	苏丹黑染色
suppository	栓剂
susceptibility of antibiotics	抗生素敏感性
suspensions	混悬剂
syrups	糖浆剂
tablets	片剂
tellurite agar	亚碲酸盐
test for sterility	无菌检查
tetracycline	四环素
t – even phage	t – 偶数序列噬菌体
tincture	酊剂
toluidine blue solution	甲苯胺蓝溶液
toothpick	牙签
transducing bacteriophage	转导噬菌体
transduction	转导
transformation	转化
transformant	转化子

treponema pauidum	密螺旋体
triple sugar iron agar (tsi)	三糖铁盐琼脂
tris – hydrochloride buffer	Tirs – HCl 缓冲液
tryptone	酪蛋白
tryptone agar	酪蛋白琼脂
tryptone broth	酪蛋白肉汤
tryptophan deaminase	色氨酸脱氨酶
tryptose	胰蛋白
tweezers	镊子
ultraviolet (UV) radiation	紫外辐射
ultra – cold frozen	超低温冷冻
urea agar	尿素琼脂
urease	脲酶
urease test agar	脲酶测试琼脂
variation	变异
virus	病毒
vaseline	凡士林
voges – proskauer test	V – P 实验
water – adsorption paper	吸水纸
wort agar	麦芽汁琼脂
wright's stain	赖氏染色
xylene	二甲苯
yeast	酵母
yeast extract	酵母膏
YPD	酵母浸出粉胨葡萄糖琼脂

参考文献（References）

[1] 周长林．微生物学实验与指导（双语）[M]．北京：中国医药科技出版社，2004.

[2] 程永宝．微生物学实验与指导 [M]．第 1 版．北京：中国医药科技出版社，1994.

[3] 周德庆．微生物学实验教程 [M]．第 2 版．北京：高等教育出版社，2006.

[4] 徐威．微生物学实验．北京：中国医药科技出版社，2004.

[5] 沈萍．微生物学实验．第 4 版．北京：高等教育出版社，2007。

[6] 黄秀梨．微生物学实验指导．第 2 版．北京：高等教育出版社，2008.

[7] 朱立平，陈学清．免疫学常用实验方法．北京：人民军医出版社，2000.

[8] 国家药典委员会．中华人民共和国药典（二部）．北京：化学工业出版社，2005.

[9] 张卓然．医学微生物实验学．北京：科学出版社，1998.

[10] John P Harley. Laboratory Exercises in Microbiology6th. McGraw – Hill College, 2004.

[11] Steve K Alexander and Dennis Strete. Microbiology: A Photographic Atlas for the Laboratory. An imprint of Addison Wesley Longman, 2001.

[12] James G Cappuccino and Natalie Sherman. Microbiology, a Laboratory Manual. Benjamin Cummings, 2002.

[13] I. Edward Alcamo. Laboratory Fundamentals of Microbiology. Jones & Bartlett Publishers, 2001.